New Developments and Environmental Applications of Drones

Tarmo Lipping • Petri Linna • Nathaniel Narra
Editors

New Developments and Environmental Applications of Drones

Proceedings of FinDrones 2020

Springer

Editors
Tarmo Lipping
Tampere University
Pori, Finland

Petri Linna
Tampere University
Pori, Finland

Nathaniel Narra
Tampere University
Pori, Finland

ISBN 978-3-030-77859-0 ISBN 978-3-030-77860-6 (eBook)
https://doi.org/10.1007/978-3-030-77860-6

This Springer imprint is published by the registered company Springer Nature Switzerland AG
The registered company address is: Gewerbestrasse 11, 6330 Cham, Switzerland

Preface

Drones or, more officially, unmanned aircraft systems (UASs) have made their way into everyday life during the past decade both as handy appliances for private use and as professional equipment fitting into the workflows of a large variety of businesses. When considering the trends in drone usage, various aspects need to be considered such as technological issues, regulations and policies, operational issues (including training), data and computation, as well as more specific issues related to particular application areas. This volume is a collection of contributions submitted to the Findrones2020 conference held virtually in November 2020 in Pori, Finland. The aim of the conference was to serve as a forum for sharing experiences and discussing future trends in the field. While the general scope of the conference was wide, including trends in drone technology, education, and regulatory issues, a special focus was on the application of UASs in agriculture, forestry, and environmental monitoring. The first part of this book presents an overview by Kramar and co-authors on the technological and operational challenges UASs face, especially when operated in harsh conditions such as the Arctic. This wide-scope review encompasses virtually all aspects of UAS usage and gives the reader a good introduction to the subject.

In the second part, chapters are divided into two categories: those related to drone technology and navigation, and those related to the usage of UASs in agriculture. In their chapter, Suurnäkki et al. describe an experimental setup, involving an icing wind tunnel, for testing the effect of ice on the dynamics of a drone propeller. They find the setup suitable for studying the challenges of Nordic conditions on drones, pointing out some needs for further development. Saffre et al. develop a methodological framework for autonomous operation of a swarm of drones for surveillance and situation awareness applications. The swarm operates autonomously on a predefined mesh. The performance of the framework is analyzed with respect to the number of drones and base placement strategy. In another chapter dealing with technical issues, Palovuori addresses a means of rendering a drone invisible against the sky by automatic control of the LED tapestry. This part of the volume is concluded by a chapter by Nevalainen et al. presenting a self-corrective simultaneous localization and mapping framework for long-term drone missions.

The considered application is in the field of forestry while the paper focuses on the correction algorithm development for map updating.

The third part of this volume begins with a review by Kaivosoja on the possibilities for UAV-based imaging in smart farming. Various aspects of drone usage such as improving target visibility, development of sensor technologies, data management, and integrating robotics for actuation are considered. In their contribution, Halla et al. present a general framework for estimating the availability of water in soil for plants using various means of data collection, including drones. The data acquired by drones can be considered as a proxy for direct measurements of water content in the soil. Linna et al. focus on the feasibility of using ground penetrating radars (GPR) in data acquisition from crop fields and on the possibility of mounting the GPR to drones. The range of possible applications is wide, ranging from soil mapping to the detection of pipes and cables. In their chapter, Narra et al. propose a cost-effective solution for the assessment of crop yield by dividing crop fields into zones according to productivity and extending the results of yield sampling to the whole field according to these zones. Obtained yield maps provide decision support to farmers when considering actions for improving the productivity of their fields. In the final contribution by Nevavuori et al., a data fusion approach is taken to improve the performance of the previously developed deep learning framework for yield prediction and modelling. A way of combining remote sensing data from drones with data on soil properties, weather conditions, and low-resolution images from satellites is studied.

The chapters contained in this volume form a tiny snapshot of the wide range of issues to be considered in drone usage, even when narrowing the scope to environmental applications and smart farming. New technological advancements and avenues for applications are reported almost daily. Developments in sensor technology widen the range of useful applications while more widespread use of drones brings the prices down. Challenges posed by unfriendly and harsh environments have to be considered through careful planning of operations as well as technology developments. As with other rapidly developing technologies, governmental regulations tend to lag and need to be updated continuously. While not making compromises on safety and security, the regulations should enable the uptake of new advancements that would lead to more optimal and environment-friendly solutions. We hope that the current volume would give the reader new ideas and promote future advancements in drone usage.

Pori, Finland Tarmo Lipping

Pori, Finland Nathaniel Narra

Pori, Finland Petri Linna

March 2021

Acknowledgments

This book was made possible by the domain expertise and efforts of the members of the organizing committee of the FinDrones 2020 conference:

- Petri Linna (Tampere University, head of the committee)
- Juha-Pekka Alanen (RoboCoast Digital Innovation Hub)
- Timo Kerminen (Satakunta University of Applied Sciences, RoboAI)
- Tinja Pitkämäki (University of Turku)
- Jere Kaivosoja (Natural Resources Institute Finland)
- Paavo Kosonen (Arctic Drone Labs)
- Roope Näsi (National Land Survey of Finland)
- Juha Tiainen (Seinäjoki University of Applied Sciences)
- Timo Lind (VTT Technical Research Centre of Finland)

We also thank the five respected high-level invited keynote speakers for their insightful and thought-provoking talks at the conference (mentioned here in order of presentation):

- Pauliina Laitinen: *"Drone based plant protection? (orig. in Finnish: Drooneilla turvallista kasvinsuojelua?)"*
- Antti Lipasti: *"Drones and 5G in Agriculture – Drones in agriculture now and in the future"*
- Jesper Cairo Westergaard: *"Drones in agricultural research and plant breeding"*
- Jonas Stjernberg: *"Scaling up advanced drone operations"*
- Pirkka Herpiö: *"Education now and in the future (orig. in Finnish: Koulutus nyt ja tulevaisuudessa)"*

In the year of COVID-19 related disruptions, the conference was able to be held virtually, thanks in large part to the efforts of staff at Satakunta University of Applied Sciences (Pori, Finland) and RoboAI Competence Center. Their assistance in setting up and operating equipment at their premises was invaluable and ensured a successful event.

We would also like to thank our conference partners—Pyhäjärvi Institute, ProAgria, Central Union of Agricultural Producers and Forest Owners (Satakunta MTK), Työtehoseura, and Robocoast Consortium—for their support.

Finally, and importantly, we would like to thank the Regional Council of Satakunta (Satakuntaliitto), through whom the PeltoAI project has been funded by the European Regional Development Fund.

Contents

Unmanned Aircraft Systems and the Nordic Challenges

Vadim Kramar, Juha Röning, Juha Erkkilä, Henry Hinkula, Tanja Kolli, and Anssi Rauhala

Abstract The European Union (EU) regulations regarding the unmanned aircraft system (UAS) that came into force in 2021 emphasise technological and operational safety. Those regulations have been developed on the common rules in the field of civil aviation and establishing a European Union Aviation Safety Agency (EASA). The implementation of the regulations and compliant UAS operator activities are still the ground of the future. Therefore, it is essential to systematically gather information about all the factors affecting UAS operations in a safe and meaningful manner. This book chapter introduces the Nordic as well as generic challenges for UAS operations. The challenges can be divided into two main categories: technological and operational. Based on the extensive literature review and authors' practical experience, both types of challenges are grouped by relevance topics. For example, the weather-related phenomena challenges overlap in both technological and operational categories but still can be mitigated differently. Technological challenges are usually mitigated by UAS design and human-computer interactions, while operational challenges may be mitigated with legislation and organisational activities and personal UAS operator qualities. Finally, the needs for further research on the challenges affecting safe UAS operations are discussed.

V. Kramar (✉) · H. Hinkula
IT Department, Oulu University of Applied Sciences, Oulu, Finland
e-mail: vadim.kramar@oamk.fi; henry.hinkula@oamk.fi

J. Röning
Faculty of Information Technology and Electrical Engineering, Biomimetics and Intelligent Systems Group, University of Oulu, Oulu, Finland
e-mail: juha.roning@oulu.fi

J. Erkkilä
Centria University of Applied Sciences, Digitalisation, Ylivieska, Finland
e-mail: juha.erkkila@centria.fi

T. Kolli · A. Rauhala
Faculty of Technology, Structures and Construction Technology, University of Oulu, Oulu, Finland
e-mail: tanja.kolli@oulu.fi; anssi.rauhala@oulu.fi

T. Lipping et al. (eds.), *New Developments and Environmental Applications of Drones*,
https://doi.org/10.1007/978-3-030-77860-6_1

Keywords UAS · UA · UAV · Drone · Arctic · Nordic

1 Introduction

Future autonomous mobile systems will employ artificial intelligence (AI) technologies to be capable of operating in an often unstructured and dynamic environment in a safe and meaningful manner and simultaneously work toward given mission objectives without being extensively controlled by human operators [1]. In this book chapter, challenges associated with the unmanned aircraft system (UAS) operations are introduced. Special attention is given to the Nordic or even the Arctic conditions' requirements on the application of UAS. The Nordic weather conditions are often harsh and include a wide range of phenomena. Simultaneously, the technology readiness level of using UAS is advancing rapidly, which has resulted in regulative actions such as the European Union (EU) regulations that came into force at the beginning of 2021. Besides, EU 2019/945 [2] and EU 2019/947 [3] crucially emphasise UAS operations' safety.

These days, the application of UAS in cold and harsh environments such as the Arctic and Antarctic is comprehensive. It includes remote sensing in fluvial environments [4], wildlife [5] and airborne [6] monitoring and population ecology [7], tracking of river ice [8] and sea ice movement [9], snow extent mapping [10], estimating the mass and body condition of animals [11], air quality measurements [12], spatial ecological and landscape surveys [13], observing the atmospheric phenomena [14], boundary layer [15] and profiling [16], monitoring changes on a construction zone [17], military purposes, e.g. [18], and many others. In addition to the professional application of UAS for research, business, and military purposes, the consumer's applications will also appear increasingly with the rapidly increasing Arctic tourism [19]. Moreover, there is a spreading area of non-commercial application of UAS for the benefits of society. With the involvement of private individuals, academic researchers, journalists, non-governmental organisations (NGOs), and sometimes public services and commercial structures, UAS may be used for humanitarian aid, environmental protection, emergency services, responsible journalism, and activism [20].

As UAS have become more popular even among regular consumers in addition to hobbyists and professionals, international or national authorities in most of the developed countries have introduced regulations on the use of UAS during the 2010s [21, 22]. However, due to the rapid emergence and development of the technology, regulations may have been straggling in some countries and thus placed barriers on research, development, market opportunities, and societal gains [21]. Moreover, the applicability, technical requirements, operational limitations, administrative procedures, and ethical constraints have varied significantly between different countries [21]. Due to the adaptation of EU level regulations [3], new administrative procedures will be implemented. For example, in Finland, the registration of unmanned aircraft (UA) is now required except for the toy UAs that weigh less than 250 g. The registration process and the knowledge exam and relevant training materials for so-called Open category applications of UAS are to be arranged and provided by the National Aviation Authority (NAA), Finnish Transport and Communication Agency (Traficom). In addition to that, NAAs empower other trustful organisations to provide Open category training and professional training for Specific or Certified categories of UAS operations. The last two categories demand from UAS operators considering more aspects in planning and conducting the UAS missions. Those aspects include safety and risks, policies and procedures, considerably air traffic rules and legal measures, infrastructure particularities, and many others [3].

An example of a Specific category training opportunity is the DroneMaster project (2020–2022) [23], aiming to develop and bring online education for professional UAS operators. The professional applications of UAS imply UA missions in a challenging environment and under critical conditions. Due to the geographical location of Finland, many professional UAS operations take place in challenging weather conditions. For that reason, the project team has collected the technological and operational requirements systematically. The DroneMaster project [23] team has conducted qualitative research using a questionnaire that has indicated how the UAS operators and end-users utilise UAS in the Northern Ostrobothnia area of Finland. From the questionnaire results, it was possible to point out that 77% of the UAS operators use commercial UAS solutions. The most notable UAS application areas in Northern Ostrobothnia are media-centric (mostly entertainment and photography related), sales and marketing, nature and animals, geology, mapping, and topography. These application areas use UAs of a variety of mission-specific payloads. According to the respondents, the Nordic conditions, especially in the winter season, and the UAS technology maturity are challenging for the UAS operators in the above and many other application areas. Today, specific technological solutions give UA's components protection from such weather conditions as rain. Still, the winter conditions will affect the UAs even stronger. The weather can also affect the mission-specific payloads' functional properties, like cameras and sensor platforms. Often, those payloads do not have any Ingress Protection (IP) [24], and therefore wind, rain, and freezing temperatures might cause their malfunctions. The questionnaire answers indicated that the UAS operators would like to have more information regarding the new European Commissions legislations (EU 2019/945 [2] and EU 2019/947 [3]), as well as modern UAS technology, maintenance aspects, and how to operate UAS.

Within this book chapter's scope, the specified in EU acronyms will primarily be used: UA, UAS, and a UAS operator. According to EU legislation [2], the definitions are given as follows:

(1) 'unmanned aircraft' ('UA') means any aircraft operating or designed to operate autonomously or to be piloted remotely without a pilot on board;
(2) 'equipment to control unmanned aircraft remotely' means any instrument, equipment, mechanism, apparatus, appurtenance, software, or accessory that is necessary for the safe operation of a UA other than a part and which is not carried on board that UA;

(3) 'unmanned aircraft system' ('UAS') means an unmanned aircraft and the equipment to control it remotely;
(4) 'unmanned aircraft system operator' (UAS operator') means any legal or natural person operating or intending to operate one or more UAS.

The following acronyms and terms are to be considered interchangeable as they all are used in reference materials: unmanned aerial vehicle (UAV), unmanned aircraft (UA), and drone. Widely spread recognition of UAV as "lightweight aircraft" is not appropriate since some UAs are heavy-duty drones that can carry a payload of 30 or more kilograms, e.g. GRIFF [25], or even a human, e.g. EHang [26]. Some as heavy as aircraft are, e.g. NATILUS [27] or Northrop Grumman MQ-4C Triton [28].

2 The Nordic Challenges

The Nordic countries have an intermediate climate with four seasons. In most Nordic countries, the summers are bright and warm, and winters are dark and cold. It is common to have heat waves during the summer season when temperatures will rise to almost 40 °C. During the winter season, the temperatures can plummet up to −40 °C, and the lowest recorded in the Arctic is −68 °C below zero. Depending on the definition, the Arctic region starts north of 60° north latitude or north of the Arctic Circle (66°33′44″ N). Already during the autumn season, there can be quite a lot of high rainfall, snowfall, and darkness since the daytime gets ever shorter as the upcoming winter Polar Night draws closer. During that time, to the north of the Arctic Circle, the sun does not raise at all. After the Polar Night has passed, the daytime grows longer until it is the longest during the midsummer when the summer solstice occurs. During that time, to the north of the Arctic Circle, the sun does not set at all.

All those factors multiply the impact of generic challenges for UAS operations and form combinations of factors that increase the overall severity significantly. For that reason, the generic UAS challenges are listed and grouped along with the specific to the North, and all together are considered. All these challenges for UAS operations are categorised into technological challenges and operational challenges [29, 30].

Technological challenges may be addressed, and their impact reduced or eliminated with the development or excellence of technologies, improved design or functionality of UAS, more sophisticated construction materials, or application of additional technological means or artefacts. Addressing the collections of challenges belonging to those topics requires different technological approaches, even though some of those approaches may be common [29, 30].

Operational challenges are those that may be addressed with human actions, predictive or corrective. Some problems or risks brought by the operational challenges may be reduced or excluded by access to information content under proper awareness and preparedness. Some operational challenges (e.g. legislative or weather-related) are not possible to overcome, and therefore, they must be taken into consideration at planning and operational phases when the decisions are to be made whether to cope with those or cancel or reschedule the UAS mission [29, 30]. Addressing the collections of challenges belonging to those topics requires the UAS operators' different types of actions, even though some of those actions may be similar or sequential.

2.1 UAS Design-Related Challenges

Different classifications of UAS exist [31, 32]. In Europe, one of the proposed categories for small UAs is by maximum take-off mass (MTOM), which is under 25 kg [2]. Light UAs are up to 150 kg, and large UAs of more than 150 kg are not classified any further. Body materials, main construction, and moving parts may be designed for a specific application or even a particular UA mission. Depending on the construction type, different design challenges appear [31]. Some of the earlier research suggests that fixed-wing lightweight UA would be suitable for a wide range of operational environments, including maritime, mountains, and arctic environments [33, 34]. Extra strong elements of UAs are produced using composites, which are very strong, stiff, and durable and at the same time lightweight (e.g. carbon fibre-reinforced composites (CRFCs)). Newly developed materials have high chemical resistance and keep their properties in a wide range of temperatures. Carbon nanotubes (CNTs) may be used to build electric-powered coating that prevents icing of UA. Composite additive-enhanced materials may be 3D printed [35].

The payload is a combination of different technological design challenges, mission-specific requirements [36, 37, 38] and design targets, e.g. to keep the MTOM within a certain range. In the specific cases of UAS applications, several additional considerations have to be taken into account while developing, e.g. transportation of vaccines requires meeting the infrastructure requirements, transportation container design, time and vibration threshold, as well as relevant to logistics economic consideration in general [39]. The payload of the UA should be known at the flight operations planning stage. The challenges are to keep the payload within the specified payload capacity of UA and achieve the physical and the data exchange specification match of the interface. Performing the UA operation with excess payload or physical mismatch may cause mission failure and safety threat. In contrast, the interface mismatch may result in malfunctions of the mission-specific payload. The general public sometimes are concerned [40] about the nature of the payload and the appearance of heavy cargo transported by UA in relative proximity. The results of the concerns may have a direct or indirect effect on the UAS mission. As the direct impact, the aggressive crowd may cause the abortion of the mission or the need to reroute the flight. As the indirect, the future missions may be suspended or cancelled.

The most critical part of the UAS that affects flight operations duration is the power source, most commonly a battery. Battery technologies are essential, regardless of UAS power sources. Power source design challenges include maximising efficiency and robustness while minimising the mass [41]. Modern power technologies such as solar [42] or fuel cell [43] may be supplemented with energy-scavenging technologies [44]. The combination of several power sources results in hybrid engines [45, 46], which are designed to improve efficiency and reliability, while the internal combustion engines are still used [46]. Even aviation gasoline-powered UA may have battery-powered ignition and computing boards. Battery technologies have developed over the years but are still prone to charge loss and voltage drop under low temperatures. Battery capacity, keeping the suitable temperature conditions, alternative energy sources, and energy harvesting are matters of cutting-edge research activities these days. These challenges [47] are addressed from both ends, by developing more sophisticated battery technology [41, 48] and by optimising the energy expenditure [49, 50]. In the case of electric-powered UAS, challenges associated with battery technologies are linked to weather physics and weather dynamics topics [33] since challenges belonging to those bring a higher energy demand.

Assistive and mission-specific sensors are a huge technological development area that progresses along with nonspecific to UAS sensor development. A vast number of sophisticated assistive and mission-specific sensors for UAS are developed worldwide. The UA-associated sensor technologies [51] may include challenges designating from a range of cameras [52, 53], or another mission-specific payload, deployment of and integration with IoT services [54], utilisation of data fusion [55], and the implementation of multi-UA setup [56], which may also be organised as UA swarms [57]. Sensor accuracy may be crucial in some application areas [12].

Computing capacity is always a compromise in autonomous systems [58] with limited capacity power sources, a wide range of sensors, intensive communications, and demand for energy-hungry data processing [31], e.g. big data processing [59]. Achieving the right balance between computing capacity, its efficient utilisation, and power consumption brings a broad range of challenges. Modern microprocessors are powerful and energy-efficient so that not only control but also some data processing and analytics functionality may be performed on board. Multilateral computation by principles of edge computing and supported with cloud services boosts computation abilities. Still, in the case of poor connectivity or its absence, unilateral computing may be beneficial. Edge computing may bring certain benefits to UAS computational architectures [60].

Operational AI aims to help with processing vast volumes of operational data to achieve higher operational efficiency. The associated challenges include designing autonomous and assisted flight control systems that may use machine vision, data fusion, machine perception, AI-enhanced communications, and data security to operate proactively, within encoded ethical principles, interact with, and provide assistance to the UAS operators through adaptive multimodal interfaces. In the future, with the help of operational AI, it will be possible to carry out fully autonomous missions [61], also life-critical [62]. Operational AI challenges are also relevant to the limitations of AI-enhanced autonomous and semi-autonomous automation and flight assistance that should reduce the cognitive workload of UAS operators [63]. These challenges are also relevant to ethical dilemmas that sometimes are not even possible to be resolved by humans, such as choosing whom to harm if no other option is available [64].

Communications expose a broad range of challenges [32, 65, 66, 67] associated with inferior quality of communication channels, absence or out-of-reach of required communication infrastructure, and insufficiency of communication parameters. These challenges play an important role, as the communications failure may cause severe consequences to the UA mission. Internet access may be absent, and even the satellite Internet connection may not always be operational in the remote Nordic and the Arctic areas. The problems with radio or satellite signals could be caused by the location, environment, or weather conditions (e.g. no connection in narrow rocky gaps). In general, any form of wireless communication is affected by those conditions and may be distorted or lost. That may affect communication for different purposes, such as vehicle-to-vehicle (V2V), swarm, air-to-ground (A2G), and command and control (C2) with Ground Control Station (GCS). In the case of radio communication, it may be affected by magnetic storms or electromagnetic interference. Some radio-frequency bands may be a subject of licensing or reserved. Certain methods aimed at improving the communications as a part of disaster management [68] may be utilised for other UAS applications in areas with poor communication infrastructure. Electromagnetic interference causes problems in communications and may cause hazards in UA electronics. Reducing or eliminating the negative impact of such a phenomenon is yet another set of design challenges. For example, electromagnetic shielding is used to protect electronics [69], but antennas are always exposable. Counter-UAS (C-UAS) technologies use the highly focused high-power electromagnetic pulse to cause hazards in electronics [70], which in the case of military UAs and those that operate in areas of intensive electromagnetic emission brings these design challenges to the high level of importance. The possible negative impacts in communications may be reduced when potential problems are known or anticipated and appropriate solutions are planned [66]. In the case of the unexpected occurrence of communication loss in the field of operations, the corrective actions may be very challenging [56].

Vertical take-off and landing (VTOL) features are usually not designed for operations in harsh environments and a moving home point, such as a boat or a ship, an iceberg, or a floating ice floe. Nevertheless, those are not rare operational conditions in the Arctic, which brings definite design challenges. The current technology development allows achieving quite challenging cases, e.g. take-off from docking stations [71] or landing on a moving platform [72]. Nordic conditions bring specific requirements for VTOL operation [31]. For example, a launching pad or base may be required, which in the case of the moving basis is a particularly challenging requirement even for the UAS operators having the appropriate skills.

A change of the air density in higher altitude is a well-known challenge by UAS operators. This challenge reduces the flight time by engaging the UA engines more aggressively to maintain the required speed and altitude [73]. The falling of UA into the air pockets may sometimes be observed visually. The front of two different atmospheric pressure areas is not possible to be recognised visually, while the information about the movement of an area with different density may be obtained from the weather forecast.

Technical malfunctioning may bring a very diverse range of challenges [32, 74], in all varieties, from power source failures through computing or communication malfunctions to UAS operators' control interface crash [75]. The severity of technical malfunctions can be partly mitigated by increasing the redundancy of mission-critical components and control surfaces [74]. Technical malfunctions may not be reserved during the mission [77], while the negative impact of those may be reduced by a fast and correct response of the UAS operators. For example, power source shortage [47, 76] may appear during a long-time expedition along with a lack of other supplies, and mitigation actions are similar. Nevertheless, in electric-powered UAs, UAS operators may utilise technological means to acquire electric power from renewable or power-generated sources. When the UA cannot return to the departure point, the search missions may be more effective if the UAS operators and supporting team are prepared in advance and backed up with positioning technologies. That kind of cases may occur when, e.g. due to technical malfunctioning, the UA parachute is engaged, or the UA is blown out of operating range by the wind. It is possible to assume that conducting the UA mission within the scope of the technical specification of the UAS in the majority of cases is not expected to cause the additional risk of technical malfunctioning [75].

2.2 *Weather- and Nature-Related Challenges*

Weather causes a lot of both technological and operational challenges to UAS operations. Challenges posed by weather physics [78] bring additional requirements that are not usually considered in the design of generic UAs intended for operations in calm weather conditions. Traditionally found in the specification of generic UAs, operational conditions limit those to the wind speed that does not exceed the speed of a UA, the operating temperature that ranges above 0 °C, and the air's humidity that does not exceed 80% or even 50%. The Specific and Certified operations of UAS may need to be performed in far more challenging weather conditions or at higher altitudes where the Open category operations are prohibited. Partially, the range of challenges caused by weather physics factors associated with the immersion of dust or moisture may be eliminated by IP protection [24]. The other, associated with icing, may be addressed with technology advances related to ice protection, de-icing, and dynamic ice accretion [79, 80, 81, 82, 83]. Some of the associated challenges may require advances in construction materials preventing condensation or routing heat from heating sources, e.g. using printed electronics [29]. Challenges posed by weather dynamics [78] by affecting the UA aerodynamic properties harm the control of UA and its battery life [47, 73]. The challenges caused by weather dynamics may exceed the UA's technical ability and therefore compromise its mission.

One of the most significant weather-related technological challenges related to Nordics and other Subarctic/Arctic locations is low temperatures, which bring a range of critical challenges affecting UA's performance [73]. Most of the consumer-grade UAs are certified to operate in a temperature range above 0 °C, and some are above −10 °C. Industry-grade UAs may be certified to operate in the temperature range above −20 °C. The most significant negative impact of a low temperature is on battery life [84, 85]. Also, sudden voltage drops may be expected (typical for lithium-ion polymer (LiPo) types of batteries) [75] and even voltage loss. Notably, that happens in the case of a change of the battery charge beyond the range controlled by an "intelligent" battery control circuit, or the battery does not have that kind of circuit. UA's high operational performance that requires high current drains the battery much faster at lower temperatures compared to the same performance within the specified temperature range above 0 °C. In the Arctic, the temperature may drop lower than −50 °C. At that low, the temperature may affect substantial parts' mechanical properties, which may become fragile. Also, liquids and lubricants' viscosity may change and therefore increase friction. Additionally, electronics' electrical properties may change due to the imbalance of control circuits and getting properties of some elements out of the operating range. Low temperature-related impacts on battery life and operation can be somewhat mitigated by using self-heating smart batteries or thermal insulation [85] of pre-heated to around +10 °C batteries [75]. Many manufacturers discourage the users from charging the batteries in temperatures below +5 °C. During the flight, temperatures can be maintained by using insulation [85] or chemical heat pads if the batteries are exposed to the elements.

Rapid temperature changes, such as rapid temperature drop, may cause the UA similar consequences as rain, mist, and fog, particularly relevant to condensation and moisture immersion. The additional requirements of these challenges are related to the UA construction materials to keep their physical properties in the broader range of temperatures, at least the anticipated ones. Furthermore, rapid temperature changes challenges may cause those weather phenomena associated with weather dynamics [73]. Temperature crossing 0 °C by changing from above 0 °C to below or the opposite way brings challenges associated with the accumulation of moisture and ice inside and on UA's outer surface. That may result in poor battery performance or failure, electronic failure, distortions of optics, and the like and have a similar negative impact as the other challenges described before in this section [73].

Freezing rain and ice fog appear at temperatures close to +10 °C and, in addition to those exposed to rain, uncover many challenges caused by instantaneously crystallising water particles. The particles may cause icing of contacts that changes electrical properties or breaks the conductivity; forms icy masses on propellers and wings that negatively affect the aerodynamic properties; or forms icy masses on a body of UA that affect negatively to the overall balance of UA in the air and may even lead to a jam of the flight control elements [73]. Freezing rain has a more substantial effect than just rain or snow since its drops immediately crystallise when they hit any surface of UA, and ice formation may grow thick. Snow brings a wide range of challenges that affect UA's flying performance, battery life, communications properties, and reception of satellite signals, just like in the case of rain [73]. Light snow may not stick to a UA surface during the flight but may stick before the land-off and after the landing. Blowing the snow may bring stacking snowflakes. Heavy snow due to a higher density of snowfall and blizzard due to a higher speed and multidirectional snow movement have more negative impacts as they directly hit the aerodynamic properties of a UA [73]. In those cases, snowflakes are more likely to stack on the UA surface during the flight. Wet snow or sleet immediately brings an additional negative impact associated with water and moisture, as well as any other melting snow does. Any snowfall negatively affects the performance of UA sensors equipped with optics.

Rain, mist, and fog bring a range of challenges associated with the negative affection on UA's flying performance, higher load to the power source, and communication abilities [73]. While rain brings masses of water and mist brings moisture, fog may cause rapid moistening and water condensation. Water or moisture may accumulate and enter the electric circuits and electronic elements. Accumulated during the flight moisture may be rich in airborne dust particles and other impurities, such as salt, which may be a reason for future corrosion and short circuits and, therefore, negatively affect UA's reliability. Wet air and drops of water on a lens may distort or make unusable UA sensors using optics. Heavy clouds negatively affect satellite signal reception, required for satellite-based positioning systems, e.g. global positioning system (GPS). Also, the clouds may make it difficult or impossible to have visual contact with UA [73]. Partially, these challenges may be addressed with advances in navigation and communications technologies [32]. A change in the air density is a challenge to the UA [73] flight performance. The lower density of the air, which typically appears at higher altitudes, negatively impacts lift and thrust forces. Especially challenging for UA may be crossing the front of two areas of different air density or falling into the air pocket. The flight-assisting automation may be programmed to reduce the negative influence of crossing the front and slipping into the air pocket. At the same time, enhanced efficiency of UA engines and aerodynamic properties improve UA's behaviour at higher altitudes [86].

The heavy and gusty wind having a speed about the speed of UA or higher may jeopardise the control and blow the UA away out of control or, in the case of autonomous operation, the operation range [73]. That may make impossible a return of UA. At higher altitudes, the wind speed may be significantly higher than at lower altitudes. It is very challenging to develop an automated piloting algorithm to compensate for heavy and gusty wind and maintain UA's plane position with the minimum course fluctuation. This challenge may be addressed by, e.g. advanced control [87] or aerodynamic models and improved control performance [88]. Wind shear, whirlwind, storms, and hailstorms [78, 73] bring a powerful combination of weather-related challenges. The relatively heavier and bigger UAs with powerful engines and excellent aerodynamics may withstand storms to a certain degree, but hailstorms bring extra factors of falling icy hails. A sophisticated flight-assistive automatic may be programmed to detect abnormal wind movement cases and further perform the corrective actions while maintaining the set altitude. Further, the flight-assistive automatic may inform the UAS operator that the UA is entering the storm area and offer the opportunity to return rapidly to the home base or the predefined point.

Dust and solid particle clouds that blow from the ground due to take-off and landing or can be brought with masses of air may temporally alter or disable visual contact and communication with UA and the performance of its optics-equipped sensors. The small particles entering to mechanical parts of UA may bring extra friction and alter mechanical performance [75]. UA flight in a cloud of tiny solid particles results in lower flying performance and shorter battery life and possible loss of visual contact and communication. UAS are used to research the distribution of ultrafine particles in the air and air quality [12, 89], while dust and solid particles in the air may also damage UA sensors. Additional protection may be needed to address these challenges [90]. Solid particle and liquid ingress protection determine two important yet essential properties of environment tolerance of UA. The protection levels are known as IP codes or International Protection Marking, which are classified by IEC 60529 standard [24]. It is vital to notice, though, that not only UA but also on-board and gimbal-attached equipment would have appropriate protection. The higher level of protection means a better tolerance, but it concerns the specified property only and does not cover a combination of external factors affecting UA's performance. For example, heavy rain or dust storm may not break the UA's seal but jeopardise its operation by bringing conditions that exceed the capacity of the UA aerodynamic properties [73].

Extreme light conditions are always challenging for optical sensors. The conditions are typical in the Nordic and even more exposed in the Arctic. During the cloudy days, harsh weather, twilight and night-time, there may be low light or no light conditions at all, e.g. during the Polar Night when the sky is closed by heavy clouds. There is an excess of light during bright sunny days, especially in the winter when the sunlight is reflected from horizon-to-horizon snow and ice masses. These challenging conditions bring special requirements for any optical equipment used with UAs, mainly their exposure and white balance properties and abilities. The machine vision technologies may be enhanced through more sophisticated sensor technologies and combinations of sensors [51, 91, 92].

The weather-related operational challenges may have a crucial impact on the UA mission. The challenges are relevant to adapting UAS operations to the environment and how the UAS operators respond to environmental challenges. Excellence in technology and UAS design may reduce the negative affection of weather factors. Still, it is not possible to eliminate those. Alternative plans, risk avoidance, rescheduling, and other mitigation activities may reduce negative affection. It is always recommended to consult the weather forecast [78] when planning the mission, while in the case of utilisation of the UTM system, the real-time weather data may be supplied. Nevertheless, under conditions of high technological readiness of UAS to withstand the weather challenges, appropriate qualification of UAS operators [64], and according to procedures allowing Specific or Certified category of operations in the different challenging environment, those operations may be performed when they are authorised [3].

Heavy and gusty wind may appear unexpectedly regardless of the weather forecast and challenge aerodynamic properties and flight control [83]. Those challenges may have a strong negative impact on the success of a UA mission [77]. Therefore, the UAS operators have to react fast and provide the appropriate response to compensate for the impact. The prompt responses are linked to the personal qualities of UAS operators. Wind shear, whirlwind, storms, and hailstorms are also challenging the aerodynamic properties of UA [83], sometimes so drastically that their influence exceeds the capabilities [83, 75] unless the UA is of a massive aircraft type. Often those phenomena may cause an intensive vibration of the UA or a sudden change in UA flight trajectory. The last may appear so rapidly that UAS operators cannot respond fast enough to correct the flight trajectory, which may lead to a crash [75].

In some cases, it is possible to respond with corrective action. That is linked to the personal qualities of UAS operators. The corrective actions of flight-assistive automatic may provide UAS operators with efficient help. When the phenomena are spotted, the UA may be set to the automated rapid return to the home base or the predefined point. Otherwise, the UAS operator may trigger that operation. UAS operators know well the decrease in density of air with higher altitude as operational challenges reducing the flight time by engaging more aggressively the UA engines to maintain the required speed and altitude [73]. Dust and solid particles may bring ground-associated and air-associated operational challenges. The more specific take-off procedures are recommended such as using a launching pad or station, which may require additional training of UAS operators. In the air, the ultra-small particles may not be seen and therefore not spotted by the UAS operators. The relatively heavy particles may move in the air in a form of a cloud. The cloud may be spotted visually, and it is generally recommended to avoid crossing its path or entering the cloud with UA that the aerodynamic properties would not be challenged [75, 83].

Short flight time is one of the most important operational challenges [75]. In the Arctic, territories are large, but at the same time, many places are difficult to reach. The flight time of UA may be affected negatively by harsh and even more by extreme weather conditions. Therefore, UA missions must be well-planned, and during the mission, operators should provide responsive and even proactive control. Also, this challenge is relevant to the availability of supply [42, 83].

Similarly, extreme light conditions challenge operational and mission-specific US sensors, which affects UAS operators' performance relying on visual data from those sensors or the quality of data collected with the help of those sensors [75]. The negative effect may be reduced with sensor modifications or special tactics that consider the light source position [75]. Depending on the extreme, the UAS operators and the supporting team may require auxiliary light sources or direct sunlight protection. Navigation systems may be impacted by heavy clouds (satellite) or extreme light conditions (visual sensor-based). To avoid the presented above challenges, performing the missions during the daytime in cloud-free conditions is recommended [93]. Other impact factors presented earlier [29], and those that are specified by mission requirements, may demand more sophisticated UA technologies utilising multi-sensor non-satellite positioning.

The nature-related operational challenges are relevant to all the other elements of nature but weather. Those elements are geographical irregularities, elements, and formations, as well as wildlife, fish, and plants. Geographical irregularities, such as cliffs and crevices, have a fixed position, while piles of ice and icebergs are moving obstacles [29]. The associated challenges are related to selecting the home point, the mobility of the UAS operators, or the UA flight path [75]. Operating UA at a low altitude above the open water with floating ice floes or during the ice breaks may bring such unexpected obstacle as suddenly pushed up ice floe [29]. Possible blind areas due to geographical formations may prevent communications between the UAS operator and the UA [94]. It is recommended to be aware of local geographical irregularities and prepared for corrective actions [30, 75]. Wildlife may be disturbed by UA and expose a wide range of emotions from internal stress to angriness [95, 96]. The challenge is that the UA may be attacked by wildlife [97] or even the UAS operator, especially when their intrusion is considered as a threat to the offspring. In addition to that, the UA performing a flight may be accidentally damaged by fighting animals. It is recommended to get acquainted with the presence of wildlife in the area of the UAS mission and minimise the possible negative impact by avoiding to disturb it [95, 96].

2.3 *Legislation and Organisational Challenges*

According to formalised or freely planned routines, challenges posed by legislation and organisational actions frame the UAS operators' efforts. The review of such challenges is following.

Best practice-associated challenges may be technology-driven or generic. The current challenges are in following the best practices, recreating their routines, and conducting expected actions or responses to the field's situations [75]. The technological challenges have already been researched for several application domains. The application domains include environmental monitoring [98, 99], forestry applications [100, 101], wildlife monitoring [53], disaster management [68], hydrology [102], fishery science [103], radiometry [104], and soil carbon mapping [105]. Many of those practices include generic operational aspects as well. Nevertheless, generic [106] or generic but still more specific to the Arctic [107] collections of best practices exist.

Laws, policies, and regulations are countries or geographic area dependent. Legislative materials may outline restrictions and prohibitions, describe operational practices, and contain information regarding required processes, licences, and permissions. All those matters bring associated challenges to UAS operators [75]. Among those to be considered are challenges relevant to generic or Open applications of UAS and relevant to the Specific and Certified operations. For example, health or medical applications of UAS [77] may require certification of mission-specific equipment.

Any of the UAS missions should not violate human rights, but the general public has many concerns conditioned by sometimes unethical applications of UAS [108]. The associated challenges are avoiding erosion of human rights, which has to be considered at the entire UAS operations process from the planning stages to the mission outcome processing [109].

Ethics and privacy considerations are already in the focus of legislative regulations in different countries but are still be researched further [77]. The main challenge for UAS operators is to be aware of those considerations, bear those in mind, and act appropriately while operating the UA [64, 75]. Heightened privacy considerations, including data protection, concerning both the society members and UAS operators, are a subject for the general public and policymakers [109].

2.4 *Challenges of Human-Computer Interaction and Personal Qualities of UAS Operations*

Challenges posed by UAS operators and other human actors' personal qualities are mostly due to the combination of knowledge, experience, and individual perception of those. Nevertheless, a range of personal qualities, including emotional intelligence, moral, quick decision-making, and fast motoric, are essential for UAS operators. Those qualities can be developed as a result of training, coaching, or extensive practical experience. The review of the associated challenges is following.

Human-computer interactions (HCI) related operational challenges are causally linked with their technological counterparts, but they expose the UAS operator's view on and reflection to the technological limitations or excellence. Any challenge of this type may complicate or prevent the UAS mission, result in mission failure, and cause safety risks. The review of such challenges is following.

Control interfaces may help to eliminate human limitations such as slow reaction compared to a real-time control system and inability to pay attention to several objects under control [110] and assist in decision-making processes [111]. Human-robot interactions have been researched well before the UAS era [112]. Similar principles apply and may be enhanced with assistive technologies [113]. Achieving that brings a broad range of design challenges, typically associated with human-computer interactions, and in the case of intelligently enhanced interactions is linked with operational AI challenges. Innovations relevant to UAS control interfaces range from direct controlling, e.g. using gestures, 3D interfaces or video streaming systems enhanced with the first-person view (FPV). More specific control interface requirements may come from the UAS application domain, e.g. infrastructure inspection [114] and civil engineering [115]. The associated challenges include the UAS operators' ability to use those interfaces and benefit from their power. Modern concepts, such as cognitive human-machine interfaces and interactions (CHMI2) [116], bring UAS operators practices to the new level but may require additional training.

Infrastructure requirements are sometimes critical for UA mission. They may not be achievable in a particular geographical area to match the demand for a specific UAS mission, e.g. duration of the expedition [75, 76]. The occurrence of such challenges may have a dramatic effect on the UAS mission. For example, it may require engaging additional technological solutions, e.g. non-GPS navigation or the communication channel must be established, or deployment of additional equipment, e.g. a portable base station or alternating current (AC) power outlet arranged for C2 equipment. It is recommended to perform a take-off from a natural or artificial object standing at a certain height and a reasonable distance from the snow to reduce blown snow's negative impact. Adapting to a different temperature (e.g. when UAS is taken indoor or outdoor) is recommended to be performed slowly, in a lengthy period and using a buffer zone. Some of these infrastructure elements may not be available in the Arctic fields, or there may be no infrastructure available at all [75]. This challenge may be addressed with a mission-specific temporal infrastructure set that may be deployed for the UAS mission [117]. A similar approach can be adapted to the other UAS operations in the field. Organising the UAS mission in the wild may require compromises between the required and the available infrastructure, and therefore a more thorough mission planning is needed. For this type of case, it is recommended to conduct a cost-benefit analysis to evaluate UAS applications' effectiveness [77].

UAS traffic management (UTM) systems raise combined technological challenges linked to communications, computing capacity, control interfaces, infrastructure requirements, just-in-time/dynamic data supply, safety, security, technical malfunctioning, and time constraints [118]. The combined challenges are valid to any geographical area where the UTM services are provisioned [119, 120, 121]. UTM-associated challenges may appear when dynamically updated information due to newly discovered circumstances is such that may require re-routing or may affect the UAS mission dramatically. Then, the entire support team's involvement may be needed for rapid decision-making [122].

Beyond visual line of sight (BVLoS) missions bring a broad range of technological challenges associated with communications, remote controlling, rerouting, and other operations of autonomous or assisted navigation, localisation, operational AI, data protection, and cybersecurity [74, 123, 124, 125]. In the case of multiple UA missions [119], like swarms of UAs, the challenge is even more severe [65]. Besides the organisational challenges, BVLoS may require the authorisation of the mission and UAS certification. It might also be a subject of UTM systems. The recently published UAS regulations in Europe categorise such missions as a Specific or Certified category [2, 3]. The rigid guidelines restricting BVLoS UAS missions in different countries have to be checked [77] to comply with the most recent regulations.

The weight of UAS and auxiliary equipment required for a mission may be important. Sometimes, the mission must be performed by one UAS operator only.

Therefore, the complete set of equipment must be lightweight and portable enough to be carried or moved by one person. On the other hand, too lightweight UA may not have enough capacity to withstand wind power and not being blown away. Industry-grade UAS sometimes require a team that deploys a launching base and operates the mission. A reasonable balance of equipment mobility should not require having an extra team member in addition to the operational team [75]. Essentially, the weight of empty UA itself is a part of take-off mass and therefore is in relevance to the operational time and payload [31, 67, 68]. Evaluating the ability to reach the operation point is especially important when the UAS operators and the supporting team move to remotely located areas to deploy the temporal home base. Cost-benefit analysis may help with precise planning to optimise the expenses and evaluate the UAS mission's effectiveness [77]. These challenges may be positively affected by novel constructive materials such as carbon nanotubes [79].

Time constraints are an essential consideration and sometimes an extreme challenge in rapidly changing weather conditions. Those may make it impossible to perform the planned mission or conclude the ongoing. Even more challenging, life-critical missions require proactive planning and fast reaction to the fast-changing situation. For example, optimal placement of UASs to cover the operation area, optimal routing, and help with decision-making may significantly help to cope with time constraints [126, 127, 128, 129]. Time constraints are affected by take-off and landing time and the speed of UAS. The main associated challenges are linked to battery technologies, payload, and fuel type. When UASs are utilised in life-critical applications, time constraints are among the crucial factor of the mission success [126, 127, 128, 129].

Coordination with professional operations (e.g. search and rescue, military, police, medical, etc.) is required for harm- and hampering-free missions [109]. In some cases, the challenges of coordination may be faced for elimination by all the involved parties. The co-involvement may be achieved (e.g. in the case of search and rescue and backcountry medical response [130]) under conditions of UA's technical readiness, a qualified UAS operator, compliance with policies and regulations, and whenever needed infrastructure support. That may imply interoperability with or integration into the professional information system, including UTM.

Following guidelines and professional codes of practice is a high goal [131], and achieving it is a different challenge for UAS operators depending on their moral qualities [64]. Consulting existing guidelines and professional codes of practice should be essential for all the stakeholders and actors of UAS operations [109]. Planning and following the plan is often challenging in the case of any type of human activities. In extreme, rapidly changing conditions, following the procedure strictly is not always feasible and safe. Sometimes, advancing or preventive action or even cancellation of a mission is necessary. The best plan, though, is where all the possible changes are anticipated and considered in advance [75]. UAS mission following the plan is essential unless there is a strong need to either change to the backup plan or act proactively [132].

Human factors are often more critical to the success of a mission than technical excellence [124]. Among many others, the associated challenges include the level of competence, the sum of experience, the degree of understanding UAS and on-board and auxiliary equipment and their technical abilities and limitations, and the ability to proact or react to changing circumstances. The challenges exposed by human factors are difficult to predict, but when they appear, their negative impact may be reduced with classical approaches for human-robot interactions [112] and preventive awareness [111]. Human responses, challenging the UAS operations in general or particular mission, public or personal, may vary significantly [75]. Adverse psychological and physiological human reactions to UAs still require more detailed research [109]. It is recommended to anticipate or be aware of possible negative responses, and in some cases, the decision is between to continue the UA mission or cancel it [40, 133]. UAS operators' insufficient qualification may not be a reason for mission failure but may bring unwanted risks [124]. Therefore, for some specific missions, additional training and the official certification of UAS operators are required [3]. Emotional aspects challenge the UAS operators, the supporting team, and the other actors of UAS missions, including the object of the mission, e.g. the victim to whom the life-critical UAS mission is aimed [129]. The personal quality of UAS operators and their experience help them withstand the influencing factors. Inter-communications and support for decision-making may also help all the actors of the mission, even in the case of automated missions [129].

Formal procedures, e.g. for operation and maintenance [51], may provide more streamline experiences with a smaller probability for errors. The UAS operator-associated challenges are to know and follow those procedures. Along with guidelines and professional codes [131], all the procedures may still not be formalised in the nearest time or not formalised at all. Some effort towards formalisation has already been taken in Europe [3].

Process-relevant challenges [51, 40, 131] occur when generic, typical, or normalised processes are not sufficient enough, and a deviation or even breakthrough is required. Such situations may be passively adopted or actively developed. An example of the first is the need to use extra solid gloves to control UA on a frosty day. The second is essential for research. For example, a commercial off-the-shelf (COTS) consumer-grade UAS is used for non-trivial tasks for which it has not been designed. That may require additional equipment or improvement of UAS itself or the rationalised operational process. These types of challenges are usually known in advance and may be considered during the planning phase.

2.5 *Infrastructure-Related Challenges*

Infrastructure-related challenges may cause issues regardless of the excellence of the UA design, even though technological excellence may reduce the negative impact. The review of such challenges is following.

AI data post-processing is widely used whenever volumes of data, requirements for processing speed, other demands, or processing-specific requirements do not match with the ability of processing by humans. One of the typical AI application areas for data post-processing is 3D mapping [134] and other photogrammetry tasks [98]. A considerable number of UAS applications relevant to monitoring, mapping, and observation require data post-processing, e.g. forestry [100]. The associated challenges include such considerations as a format of data, compressing and encryption, storing data on board, or transferring to the Ground Control Station (GCS), which is linked with the communications challenges. With the current development of UA on-board computing capacity, it is possible to build 3D models of the environment in real time to support operational AI's functioning, e.g. for navigation purposes [135].

Satellite-based navigation systems may be altered or not always be available at high north latitude. Harsh weather conditions can aggravate the situation. A magnetic compass of UA may be disoriented and require frequent recalibration. Gyroscopic compass may not be portable enough for a lightweight UA, and it may also be compromised. Compass problems are conditioned by a Magnetic North Pole shift versus the Geographic North Pole, terrestrial magnetic distribution in the Arctic, and magnetic aberrations (also relevant to solar activity). A lack of precise location has a direct negative impact on an automated routing and assisted control and the outcome of UAS mission or quality of gathered data, for example, in the case of image co-registering. Some of these challenges may be addressed with a variety of technology advances [31, 62, 115, 135, 136, 137] and combined with geofencing technologies enhanced with satellite remote sensing [138].

Safety is one of the key set of technological challenges [139, 140]. A variety of safety-relevant aspects are investigated and addressed [67, 124]. UAS provide enormous potential safety, e.g. in search and rescue, disaster assessment and response, and hazard monitoring [94], but there are obvious concerns regarding, e.g. aviation safety and malicious use of UAS [139]. Safety is always to be considered by UAS operators as the ultimate goal of any UA mission. The safety considerations and associated challenges for the UAS operators are broad. They include public safety [67]; the physical safety of involved human actors and wildlife; the safety of scientific data, equipment, and infrastructure; and air space safety in general [75]. Following the safety regulations, formal procedures, and best practices, thorough planning of the mission, taking care of the technical condition of UAS, operating carefully, and providing a prompt response in case of extraordinary situations contribute to the safety [38, 75, 124].

Security is another crucial set of technological challenges that may be addressed in a variety of ways [32, 40, 139, 141, 142]. Possible security threats include, e.g. jamming or spoofing the localisation data or UAS transmissions, manipulation of captured footage, injection of falsified sensor data, malicious software, denial-of-service (DoS) attacks, and GCS control signal jamming or spoofing [142]. The security breach is one of the severe challenges that may lead to lost data or control of UA and make the mission compromised, e.g. the flight may be rerouted [77]. Addressing possible challenges requires UAS operators to know the challenges and the ability to recognise those and respond appropriately [94, 142]. There have also been concerns of possible data capture and transmission back to the UAS manufacturer or a third party, resulting in the blacklisting of certain manufacturers from governmental applications in, e.g. the United States. However, independent audits have not validated the concerns about unwanted data transmissions, although particular vulnerabilities exist [143].

Just-in-time/dynamic data supply is crucial for data-critical missions and autonomous or assisted navigation [32, 47]. The full-duplex C2 communications with GCS may have different latency requirements depending on the nature of the UAS mission. Regardless of the requirements, communication reliability should not be compromised [65, 66]. The challenges occur when the communication channel has been compromised and the expected data exchange has fallen [47]. These challenges are also relevant to data that is to be supplied by services provisioned through UAS traffic management (UTM) systems for UAS operations in the urban environment.

Lack of supply may have a dramatic impact on the UAS mission and may cause many consequential challenges unless eliminated with preparatory and proactive actions [75]. During the mission, it is possible to run out of energy (e.g. battery charge or fuel), spare parts and tools, life support for the team (e.g. food or water), and other supplies due to a variety of reasons such as the wrong estimate of required quantity/amount or need to have, broken equipment, possibilities to lose equipment, etc. The nearest source of supply may be located unreachably far, and transportation is difficult to impossible to arrange. Such a situation may lead to a cancellation of the mission and even calling a rescue. The lack of supply may appear during local missions but have a higher impact during a long-time expedition [47, 76]. The probability of occurring may be reduced by careful planning and availability of spare equipment and a reserve of supply.

Low-carbon operation challenges are relevant to optimising the energy expenditure [49, 50], the flight path optimisation [49, 141], more efficient communications [67, 144], and the use of renewable energy sources [42, 44, 102]. Even though these challenges may be addressed with technological excellence, as soon as the low-carbon operation range beyond the single flight, other considerations appear, including integration into the supply chain, business, and operational processes [145].

3 Discussion

At the beginning of 2021, the European Union Aviation Safety Agency (EASA) published Easy Access Rules for Unmanned Aircraft Systems (UAS) [146] that contain the rules and procedures for the operation of unmanned aircraft (UA) adapted for better understanding of recent EU regulations (EU 2019/945 [2] and EU 2019/947 [3]) by the general public. For many defined types of UAS operations, it is required to evaluate the environmental and weather conditions and, in some instances, even acquire advanced weather information. Still, the collection of weather effects on the UA is to be developed further. The specific challenges and special conditions that affect the UA missions in the Nordic environment or at the higher altitudes of the mountain areas are not yet considered and reflected by the existing regulations.

More research is needed to investigate further the Nordic challenges and their particularities from the technological perspective. For example, such technological challenge as safety may be addressed by equipping the UA with the parachute system and duplicating its power and electric circuits, engines, and propulsor system. However, most of those safety systems are typically designed to operate in mild environmental conditions. Some of the weather challenges typical to the Nordic though may appear in many other climate zones and, therefore, compromise the safety systems' functioning. At the same time, attention is not always paid to the security of UAS, while security breach may eliminate the excellence of the safety systems and cause unwanted consequences. The holistic view at the UAS as a technology artefact exposed to the broad set of challenges may help understand the common causes of those and interrelationships between them. That may advance technological elements so that the design of the entire UAS would achieve a higher degree of robustness and resilience to severe weather conditions and therefore achieve a higher level of safety.

Every challenge mentioned in this chapter has from several to a vast number of possible solutions. For example, anti-icing solutions include sprays, liquids, and nano-coating and composite materials. Many universities and companies work on any of those solutions around the world. Because of that, a significant number of publications exist about every solution. It is well beyond this chapter's scope to present all the possible solutions or present one state-of-the-art in-depth solution.

More systematic research on operational best practices is needed and a higher effort to popularise operational practices. Coping with operational challenges is in line with the development of UAS and supportive infrastructure technologies. Real-world challenges that humanity meets, such as the COVID-19 pandemic, accelerate applications of UAS [145]. Contribution to the professional development of the UAS operators is needed for a better match of the technological progress with labour forces for society's benefits. That includes formal education, acquiring and sharing experiences, gathering, processing operational data, and raising public awareness about UAS applications' benefits. Informative communication with the general public and discussing concerns may eliminate uncertainty and reduce unacceptance. Supportive regulations and positive perception of the general public will positively impact the development of UAS applications.

To address some of the operational challenges, raise awareness about the possible pitfalls, reduce potential risks, and simplify planning, the dedicated UAS operator's handbooks were developed [106, 107]. Also, best practices to minimise the UA disturbance to wildlife are published [147]. Nevertheless, all those guides are considered very primary.

The more formalised approach has been taken by the ASTM (formerly known as American Society for Testing and Materials) Committee that started working on New Practice for General Operations Manual for Professional Operator of

Light Unmanned Aircraft Systems (UAS) [148]. In Europe, Joint Authorities for Rulemaking of Unmanned Systems (JARUS) is working on JARUS guidelines on Specific Operations Risk Assessment (SORA) [149]. The Easy Access Rules for Unmanned Aircraft Systems explains how SORA may be implemented in practice [146]. All these documents are still under development. The symbioses of regulatory and standardisation effort, technological excellence, and operational practices for risk-free UAS operations are yet to be achieved.

4 Conclusion

The rapid emergence and development of unmanned aircraft systems (UAS) and their recent advances offer enormous potential for commercial and non-commercial applications. Still, it has also raised a broad range of technical and operational challenges. All the technical and operational challenges introduced in this chapter are typical to applications of UAS in the Nordic or even Arctic environment. At the same time, they may also be specific to other climate zones. Nevertheless, the Nordic environment reveals conditions that bring the challenges associated with UAS operations to the highest level of toughness.

The safety of UAS operations has been considered a critical measure. Environmental conditions associated with the weather and geographical and other environmental particularities bring a broad range of challenges that have a strong influence on the safety of UAS operations. The majority of UAS operations happen in mild conditions that do not set the extreme demand. Achieving a higher level of safety requires investment, but UA's crash might be even more expensive. The crash is associated with the damage of property or wildlife and may cause a risk to people's lives. What is more, it may cause reputational losses of both the UAS operator that controlled the crashed UA and the UAS vendor that produced the crashed UA.

Even though a lot of research attention is paid to specific technologies associated with the introduced challenges, there is still a lack of research to understand all the possible challenges and develop a holistic view of how the entire range of challenges impacts the overall design of UAS. Robust all-weather-resilient UAS technologies able to withstand the severe Nordic conditions could establish the benchmark.

Acknowledgement The authors would like to express gratitude to the DroneMaster project (financed by European Social Fund (ESF) S21923) [23], members of Arctic Drone Labs innovation ecosystem [150], and international partners for supplying materials, giving constructive critics, and using authors' attention to a range of projects and research experiments relevant to the UAS applications in the Nordic and the Arctic environment.

References

1. Vainio, M., Ruotsalainen, L., Banda, O.V., Röning, J., Laitinen, J., Boutellier, J., Koskinen, S., Peussa, P., Shamsuzzoha, A., Toroody, A.B., Kramar, V., Visala, A., Ghabcheloo, R., Huhtala, K., & Alagirisamy, R. (2020). *Safety challenges of autonomous mobile systems in dynamic unstructured environments: Situational awareness, decision-making, autonomous navigation, & human-machine interface*. RAAS Rethinking Autonomy And Safety Situational Awareness, Autonomous Navigation and Intelligent Control Research Task Force.
2. European Commission. (2019). *Commission delegated regulation (EU) 2019/945 of 12 March 2019 on unmanned aircraft systems and on third-country operators of unmanned aircraft systems*. European Comission.
3. European Commission. (2019). *Commission implementing regulation (EU) 2019/947 of 24 May 2019 on the rules and procedures for the operation of unmanned aircraft*. European Commission.
4. Rhee, D. S., Do Kim, Y., Kang, B., & Kim, D. (2017). Applications of unmanned aerial vehicles in fluvial remote sensing: An overview of recent achievements. *KSCE Journal of Civil Engineering, 22*, 588–602.
5. Linchant, J., Lisein, J., Semeki, J., Lejeune, P., & Vermeulen, C. (2015). Are unmanned aircraft systems (UASs) the future of wildlife monitoring? A review of accomplishments and challenges. *Mammal Review, 45*, 239–252.
6. Finnish Environment Institute & gt; Airborne Monitoring Tools for Arctic and Baltic Sea Environment (UAV-ARCTIC). http://www.syke.fi/projects/uavarctic. Last accessed 2018/05/25.
7. Goebel, M. E., Perryman, W. L., Hinke, J. T., Krause, D. J., Hann, N. A., Gardner, S., & LeRoi, D. J. (2015). A small unmanned aerial system for estimating abundance and size of Antarctic predators. *Polar Biology, 38*, 619–630. https://doi.org/10.1007/s00300-014-1625-4.
8. Alfredsen, K. (2018). Brief communication: Mapping river ice using drones and structure from motion. *The Cryosphere, 12*, 627–633.
9. Leira, F. S., Johansen, T. A., & Fossen, T. I. (2017). A UAV ice tracking framework for autonomous sea ice management. In *2017 international conference on unmanned aircraft systems (ICUAS)* (pp. 581–590). IEEE. https://doi.org/10.1109/ICUAS.2017.7991435.
10. Niedzielski, T. (2018). Automated snow extent mapping based on orthophoto images from unmanned aerial vehicles. *Pure and Applied Geophysics, 175*, 1–18.
11. Krause, D. J., Hinke, J. T., Perryman, W. L., Goebel, M. E., & LeRoi, D. J. (2017). An accurate and adaptable photogrammetric approach for estimating the mass and body condition of pinnipeds using an unmanned aerial system. *PLoS One, 12*, e0187465. https://doi.org/10.1371/journal.pone.0187465.
12. Villa, T. (2016). An overview of small unmanned aerial vehicles for air quality measurements: Present applications and future prospectives. *Sensors, 16*, 1072.
13. Bollard-Breen, B. (2015). Application of an unmanned aerial vehicle in spatial mapping of terrestrial biology and human disturbance in the McMurdo Dry Valleys, East Antarctica. *Polar Biology, 38*, 573. https://doi.org/10.1007/s00300-014-1586-7.
14. Houston, A. L. (2012). The collaborative Colorado-Nebraska unmanned aircraft system experiment. *Bulletin of the American Meteorological Society, 93*, 6.
15. Jonassen, M. (2015). Application of remotely piloted aircraft systems in observing the atmospheric boundary layer over Antarctic Sea ice in winter. *Polar Research, 34*, 25651.
16. Mayer, S. (2012). Atmospheric profiling with the UAS SUMO: A new perspective for the evaluation of fine-scale atmospheric models. *Meteorology and Atmospheric Physics, 116*, 15–26.
17. Aicardi, I., Nyapwere, N., Nex, F., Gerke, M., Lingua, A., & Koeva, M. (2016). Co-registration of multitemporal uav image datasets for monitoring applications: A new approach. *The International Archives of the Photogrammetry, Remote Sensing and Spatial Information Sciences, xli-b1*, 757–763.

18. Schaub, G. (2015). But who's flying the plane? Integrating UAVs into the Canadian and Danish armed forces. *International Journal, 70*, 250–267.
19. Halliday, W. D. (2018). Tourist vessel traffic in important whale areas in the western Canadian Arctic: Risks and possible management solutions. *Marine Policy, 97*, 72–81.
20. Wynsberghe, A., Soesilo, D., Kristen, T., & Sharkey, N. (2018). *Drones in the service of society*. A Foundation for Responsible Robotics.
21. Stöcker, C., Bennett, R., Nex, F., Gerke, M., & Zevenbergen, J. (2017). Review of the current state of UAV regulations. *Remote Sensing, 9*, 33–35. https://doi.org/10.3390/rs9050459.
22. Cracknell, A. P. (2017). UAVs: Regulations and law enforcement. *International Journal of Remote Sensing, 38*, 3054–3067. https://doi.org/10.1080/01431161.2017.1302115.
23. DroneMaster – Modular Training for Drone Operators. http://www.dronemaster.fi/en/front-page/. Last accessed 2020/12/15.
24. IEC 60529:1989+AMD1:1999+AMD2:2013 CSV | IEC Webstore | water management, smart city, rural electrification. https://webstore.iec.ch/publication/2452. Last accessed 2016/02/11.
25. Fly with a legend | GRIFF Aviation. http://griffaviation.com/. Last accessed 2019/08/21.
26. EHANG Announces $10 million series a round led by GGV Capital to take personal drones mainstream. *PR Newswire* 2014.
27. NATILUS. http://www.natilus.co/. Last accessed 2019/08/21.
28. Autonomous Systems – Northrop Grumman. https://www.northropgrumman.com/Capabilities/AutonomousSystems/Pages/default.aspx?utm_source=PrintAd&utm_medium=Redirect&utm_campaign=AutonomousSystems_Redirect#Technology. Last accessed 2018/08/21.
29. Kramar, V., & Määttä, H. (2018). UAV Arctic challenges and the first step: Printed temperature sensor. In *Proceedings of the 23rd conference of FRUCT association* (pp. 483–490).
30. Kramar, V. (2019). UAS (drone) Arctic challenges – Next steps. In *Proceedings of the 25th conference of FRUCT association* (pp. 507–514).
31. Hassanalian, M., & Abdelkefi, A. (2017). Classifications, applications, and design challenges of drones: A review. *Progress in Aerospace Science, 91*, 99–131. https://doi.org/10.1016/j.paerosci.2017.04.003.
32. Shakhatreh, H., Sawalmeh, A. H., Al-Fuqaha, A., Dou, Z., Almaita, E., Khalil, I., Othman, N. S., Khreishah, A., & Guizani, M. (2019). Unmanned aerial vehicles (UAVs): A survey on civil applications and key research challenges. *IEEE Access, 7*, 48572–48634. https://doi.org/10.1109/ACCESS.2019.2909530.
33. Funaki, M., & Hirasawa, N. (2008). Outline of a small unmanned aerial vehicle (ant-plane) designed for Antarctic research. *Polar Science, 2*, 129–142. https://doi.org/10.1016/j.polar.2008.05.002.
34. Mueller, T. J. (1999). Aerodynamic measurements at low Reynolds numbers for fixed wing micro-air vehicles. *Mech. Eng.*, 1–32.
35. Legault, M. Drones: Composite UAVs take flight. https://www.compositesworld.com/articles/drones-composite-uavs-take-flight. Last accessed 2018/05/18.
36. Herwitz, S. R., Johnson, L. F., Dunagan, S. E., Higgins, R. G., Sullivan, D. V., Zheng, J., Lobitz, B. M., Leung, J. G., Gallmeyer, B. A., Aoyagi, M., Slye, R. E., & Brass, J. A. (2004). Imaging from an unmanned aerial vehicle: Agricultural surveillance and decision support. *Computers and Electronics in Agriculture, 44*, 49–61. https://doi.org/10.1016/j.compag.2004.02.006.
37. Hadi, G. S., Varianto, R., Trilaksono, B. R., & Budiyono, A. (2015). Autonomous UAV system development for payload dropping mission. *Journal of Instrumentation, Automation and Systems, 1*, 72–77. https://doi.org/10.21535/jias.v1i2.158.
38. Kellermann, R., Biehle, T., & Fischer, L. (2020). Drones for parcel and passenger transportation: A literature review. *Transportation Research Interdisciplinary Perspectives, 4*, 100088. https://doi.org/10.1016/j.trip.2019.100088.

39. Haidari, L. A., Brown, S. T., Ferguson, M., Bancroft, E., Spiker, M., Wilcox, A., Ambikapathi, R., Sampath, V., Connor, D. L., & Lee, B. Y. (2016). The economic and operational value of using drones to transport vaccines. *Vaccine, 34*, 4062–4067. https://doi.org/10.1016/j.vaccine.2016.06.022.
40. Rao, B., Gopi, A. G., & Maione, R. (2016). The societal impact of commercial drones. *Technology in Society, 45*, 83–90. https://doi.org/10.1016/j.techsoc.2016.02.009.
41. Khofiyah, N. A., Sutopo, W., & Nugroho, B. D. A. (2019). Technical feasibility battery lithium to support unmanned aerial vehicle (UAV): A technical review. *Proceedings of the International Conference on Industrial Engineering and Operations Management, 2019*, 3591–3601.
42. Runge, H., Rack, W., & Hepperle, M. (2007). *A solar-powered HALE-UAV for Arctic research*. In: 1st CEAS European Air and Space Conference.
43. Bradley, T. H., Moffitt, B. A., Mavris, D. N., & Parekh, D. E. (2007). Development and experimental characterization of a fuel cell powered aircraft. *Journal of Power Sources, 171*, 793–801. https://doi.org/10.1016/j.jpowsour.2007.06.215.
44. Thomas, J. P., Qidwai, M. A., & Kellogg, J. C. (2006). Energy scavenging for small-scale unmanned systems. *Journal of Power Sources, 159*, 1494–1509. https://doi.org/10.1016/j.jpowsour.2005.12.084.
45. Hiserote, R., & Harmon, F. (2010). Analysis of hybrid-electric propulsion system designs for small unmanned aircraft systems. In *8th annual international energy conversion engineering conference*. Reston, VA: American Institute of Aeronautics and Astronautics. https://doi.org/10.2514/6.2010-6687.
46. Sliwinski, J., Gardi, A., Marino, M., & Sabatini, R. (2017). Hybrid-electric propulsion integration in unmanned aircraft. *Energy, 140*, 1407–1416. https://doi.org/10.1016/j.energy.2017.05.183.
47. Ader, M., & Axelsson, D. (2017). *Drones in arctic environments*. ITM.
48. Dou, X., Hasa, I., Saurel, D., Vaalma, C., Wu, L., Buchholz, D., Bresser, D., Komaba, S., & Passerini, S. (2019). Hard carbons for sodium-ion batteries: Structure, analysis, sustainability, and electrochemistry. *Materials Today, 23*, 87–104. https://doi.org/10.1016/j.mattod.2018.12.040.
49. Dorling, K., Heinrichs, J., Messier, G. G., & Magierowski, S. (2017). Vehicle routing problems for drone delivery. *IEEE Transactions on System, Man, and Cybernetics: Systems, 47*, 70–85. https://doi.org/10.1109/TSMC.2016.2582745.
50. Chiang, W. C., Li, Y., Shang, J., & Urban, T. L. (2019). Impact of drone delivery on sustainability and cost: Realizing the UAV potential through vehicle routing optimization. *Applied Energy, 242*, 1164–1175. https://doi.org/10.1016/j.apenergy.2019.03.117.
51. Pajares, G. (2015). Overview and current status of remote sensing applications based on unmanned aerial vehicles (UAVs). *Photogrammetric Engineering and Remote Sensing, 81*, 281–329. https://doi.org/10.14358/PERS.81.4.281.
52. Mohd Noor, N., Abdullah, A., & Hashim, M. (2018). Remote sensing UAV/drones and its applications for urban areas: A review. *IOP Conference Series: Earth and Environmental Science, 169*, 012003. https://doi.org/10.1088/1755-1315/169/1/012003.
53. Linchant, J., Lisein, J., Semeki, J., Lejeune, P., & Vermeulen, C. (2015). Are unmanned aircraft systems (UASs) the future of wildlife monitoring? *A review of accomplishments and challenges. Mamm. Rev., 45*, 239–252. https://doi.org/10.1111/mam.12046.
54. Hossein Motlagh, N., Taleb, T., & Arouk, O. (2016). Low-altitude unmanned aerial vehicles-based internet of things services: Comprehensive survey and future perspectives. *IEEE Internet of Things Journal, 3*, 899–922. https://doi.org/10.1109/JIOT.2016.2612119.
55. Sankey, T., Donager, J., McVay, J., & Sankey, J. B. (2017). UAV lidar and hyperspectral fusion for forest monitoring in the southwestern USA. *Remote Sensing of Environment, 195*, 30–43. https://doi.org/10.1016/j.rse.2017.04.007.
56. Yanmaz, E., Yahyanejad, S., Rinner, B., Hellwagner, H., & Bettstetter, C. (2018). Drone networks: Communications, coordination, and sensing. *Ad Hoc Networks, 68*, 1–15. https://doi.org/10.1016/j.adhoc.2017.09.001.

57. Bürkle, A., Segor, F., & Kollmann, M. (2011). Towards autonomous micro UAV swarms. *Journal of Intelligent and Robotics Systems: Theory and Applications, 61*, 339–353. https://doi.org/10.1007/s10846-010-9492-x.
58. Axisa, D., & DeFelice, T. P. (2016). Modern and prospective technologies for weather modification activities: A look at integrating unmanned aircraft systems. *Atmospheric Research, 178–179*, 114–124. https://doi.org/10.1016/j.atmosres.2016.03.005.
59. Xu, W., Zhou, H., Cheng, N., Lyu, F., Shi, W., Chen, J., & Shen, X. (2018). Internet of vehicles in big data era. *IEEE/CAA Journal of Automatica Sinica, 5*, 19–35. https://doi.org/10.1109/JAS.2017.7510736.
60. Zhou, F., Wu, Y., Hu, R. Q., & Qian, Y. (2018). Computation rate maximization in UAV-enabled wireless-powered mobile-edge computing systems. *IEEE Journal on Selected Areas in Communications, 36*, 1927–1941. https://doi.org/10.1109/JSAC.2018.2864426.
61. Faust, A., Palunko, I., Cruz, P., Fierro, R., & Tapia, L. (2017). Automated aerial suspended cargo delivery through reinforcement learning. *Artificial Intelligence, 247*, 381–398. https://doi.org/10.1016/j.artint.2014.11.009.
62. Sampedro, C., Rodriguez-Ramos, A., Bavle, H., Carrio, A., de la Puente, P., & Campoy, P. (2018). A fully-autonomous aerial robot for search and rescue applications in indoor environments using learning-based techniques. *Journal of Intelligent and Robotics Systems: Theory and Applications, 95*, 1–27. https://doi.org/10.1007/s10846-018-0898-1.
63. Hocraffer, A., & Nam, C. S. (2017). A meta-analysis of human-system interfaces in unmanned aerial vehicle (UAV) swarm management. *Applied Ergonomics, 58*, 66–80. https://doi.org/10.1016/j.apergo.2016.05.011.
64. Arkin, R. C. (2016). Ethics and autonomous systems: Perils and promises. *Proceedings of the IEEE, 104*, 1779–1781. https://doi.org/10.1109/JPROC.2016.2601162.
65. Noor, F., Khan, M. A., Al-Zahrani, A., Ullah, I., & Al-Dhlan, K. A. (2020). A review on communications perspective of flying AD-HOC networks: Key enabling wireless technologies, applications, challenges and open research topics. *Drones, 4*, 1–14. https://doi.org/10.3390/drones4040065.
66. Gupta, L., Jain, R., & Vaszkun, G. (2016). Survey of important issues in UAV communication networks. *IEEE Communication Surveys and Tutorials, 18*, 1123–1152. https://doi.org/10.1109/COMST.2015.2495297.
67. Mozaffari, M., Saad, W., Bennis, M., Nam, Y. H., & Debbah, M. (2018). A tutorial on UAVs for wireless networks: Applications, challenges, and open problems. *arXiv, 21*, 2334–2360.
68. Erdelj, M., Natalizio, E., Chowdhury, K. R., & Akyildiz, I. F. (2017). Help from the sky: Leveraging UAVs for disaster management. *IEEE Pervasive Computing, 16*, 24–32. https://doi.org/10.1109/MPRV.2017.11.
69. Horapong, K., Chandrucka, D., Montree, N., & Buaon, P. (2017). Design and use of "drone" to support the radio navigation AIDS flight inspection. *AIAA/IEEE Digital Avionics Systems Conference – Proceedings, 2017-September*. https://doi.org/10.1109/DASC.2017.8102114.
70. Lykou, G., Moustakas, D., & Gritzalis, D. (2020). *Defending airports from UAS: A survey on cyber-attacks and counter-drone sensing technologies*. Mdpi. Sensors.
71. Cocchioni, F., Pierfelice, V., Benini, A., Mancini, A., Frontoni, E., Zingaretti, P., Ippoliti, G., & Longhi, S. (2014). Unmanned ground and aerial vehicles in extended range indoor and outdoor missions. In *2014 international conference on unmanned aircraft systems ICUAS 2014 – Conference proceedings* (pp. 374–382). https://doi.org/10.1109/ICUAS.2014.6842276.
72. Herissé, B., Hamel, T., Mahony, R., & Russotto, F. X. (2012). Landing a VTOL unmanned aerial vehicle on a moving platform using optical flow. *IEEE Transactions on Robotics, 28*, 77–89. https://doi.org/10.1109/TRO.2011.2163435.
73. Ranquist, E. A., & Matthias Steiner, B. A. (2016). Exploring the range of weather impacts on UAS operations. In *18th conference on aviatation range and aerospace meteorology* (Vol. 11).

74. La Cour-Harbo, A. (2017). Quantifying risk of ground impact fatalities of power line inspection BVLOS flight with small unmanned aircraft. In *2017 international conference on unmanned aircraft systems ICUAS 2017* (pp. 1352–1360). https://doi.org/10.1109/ICUAS.2017.7991323.
75. Duffy, J. P., Cunliffe, A. M., DeBell, L., Sandbrook, C., Wich, S. A., Shutler, J. D., Myers-Smith, I. H., Varela, M. R., & Anderson, K. (2018). Location, location, location: Considerations when using lightweight drones in challenging environments. *Remote Sensing in Ecology and Conservation, 4*, 7–19. https://doi.org/10.1002/rse2.58.
76. Fornace, K. M., Drakeley, C. J., William, T., Espino, F., & Cox, J. (2014). Mapping infectious disease landscapes: Unmanned aerial vehicles and epidemiology. *Trends in Parasitology, 30*, 514–519. https://doi.org/10.1016/j.pt.2014.09.001.
77. Bhatt, K., Pourmand, A., & Sikka, N. (2018). Targeted applications of unmanned aerial vehicles (drones) in telemedicine. *Telemedicine and e-Health, 24*, 833–838. https://doi.org/10.1089/tmj.2017.0289.
78. Gultepe, I., Sharman, R., Williams, P. D., Zhou, B., Ellrod, G., Minnis, P., Trier, S., Griffin, S., Yum, S. S., Gharabaghi, B., Feltz, W., Temimi, M., Pu, Z., Storer, L. N., Kneringer, P., Weston, M. J., Chuang, H. y., Thobois, L., Dimri, A. P., Dietz, S. J., França, G. B., Almeida, M. V., & Neto, F. L. A. (2019). A review of high impact weather for aviation meteorology. *Pure and Applied Geophysics, 176*, 1869–1921. https://doi.org/10.1007/s00024-019-02168-6.
79. Gohardani, O., Elola, M. C., & Elizetxea, C. (2014). Potential and prospective implementation of carbon nanotubes on next generation aircraft and space vehicles: A review of current and expected applications in aerospace sciences. *Progress in Aerospace Science, 70*, 42–68. https://doi.org/10.1016/j.paerosci.2014.05.002.
80. Cristofaro, A., Johansen, T. A., & Aguiar, A. P. (2015). Icing detection and identification for unmanned aerial vehicles: Multiple model adaptive estimation. In *2015 European control conference ECC* (pp. 1651–1656). https://doi.org/10.1109/ECC.2015.7330774.
81. Sørensen, K. L., Helland, A. S., & Johansen, T. A. (2015). Carbon nanomaterial-based wing temperature control system for in-flight anti-icing and de-icing of unmanned aerial vehicles. In *IEEE aerospace conference proceedings 2015-June* (pp. 1–6). https://doi.org/10.1109/AERO.2015.7119206.
82. Lawson, C. P. (2006). Electrically powered ice protection systems for MALE UAVs – Requirements and integration challenges. In *ICAS-secretariat – 25th congress of the International Council of the Aeronautical Sciences 2006* (pp. 3565–3573).
83. Liu, Y., Li, L., Li, H., & Hu, H. (2018). An experimental study of surface wettability effects on dynamic ice accretion process over an UAS propeller model. *Aerospace Science and Technology, 73*, 164–172. https://doi.org/10.1016/j.ast.2017.12.003.
84. Salameh, Z. M., & Kim, B. G. (2009). Advanced lithium polymer batteries. In *2009 IEEE Power & Energy Society General Meeting* (pp. 1–5). https://doi.org/10.1109/PES.2009.5275404.
85. Cimoli, E., Marcer, M., Vandecrux, B., Bøggild, C. E., Williams, G., & Simonsen, S. B. (2017). Application of low-cost uass and digital photogrammetry for high-resolution snow depth mapping in the Arctic. *Remote Sensing, 9*, 1–29. https://doi.org/10.3390/rs9111144.
86. Paredes, J. A., Saito, C., Abarca, M., & Cuellar, F. (2017). Study of effects of high-altitude environments on multicopter and fixed-wing UAVs' energy consumption and flight time. In *IEEE international conference on automation science and engineering 2017-August* (pp. 1645–1650). https://doi.org/10.1109/COASE.2017.8256340.
87. Tafreshi, M., Shafieenejad, I., & Nikkhah, A. A. (2014). Open-loop and closed-loop optimal guidance policy for Samarai aerial vehicle with novel algorithm to consider wind effects. *International Journal of Engineering and Techincal Research, 2*, 185–191.
88. Huang, H., Hoffmann, G. M., Waslander, S. L., & Tomlin, C. J. (2009). Aerodynamics and control of autonomous quadrotor helicopters in aggressive maneuvering. In *Proceedings of the IEEE international conference on robotics and automation* (pp. 3277–3282). https://doi.org/10.1109/ROBOT.2009.5152561.

89. Altstädter, B., Platis, A., Wehner, B., Scholtz, A., Wildmann, N., Hermann, M., Käthner, R., Baars, H., Bange, J., & Lampert, A. (2015). ALADINA – An unmanned research aircraft for observing vertical and horizontal distributions of ultrafine particles within the atmospheric boundary layer. *Atmospheric Measurement Techniques, 8*, 1627–1639. https://doi.org/10.5194/amt-8-1627-2015.
90. Rüdiger, J., Tirpitz, J. L., Maarten De Moor, J., Bobrowski, N., Gutmann, A., Liuzzo, M., Ibarra, M., & Hoffmann, T. (2018). Implementation of electrochemical, optical and denuder-based sensors and sampling techniques on UAV for volcanic gas measurements: Examples from Masaya, Turrialba and Stromboli volcanoes. *Atmospheric Measurement Techniques, 11*, 2441–2457. https://doi.org/10.5194/amt-11-2441-2018.
91. Franceschini, N. (2014). Small brains, smart machines: From fly vision to robot vision and back again. *Proceedings of the IEEE, 102*, 751–781. https://doi.org/10.1109/JPROC.2014.2312916.
92. Gade, R., & Moeslund, T. B. (2014). Thermal cameras and applications: A survey. *Machine Vision and Applications, 25*, 245–262. https://doi.org/10.1007/s00138-013-0570-5.
93. Funaki, M., Higashino, S. I., Sakanaka, S., Iwata, N., Nakamura, N., Hirasawa, N., Obara, N., & Kuwabara, M. (2014). Small unmanned aerial vehicles for aeromagnetic surveys and their flights in the South Shetland Islands, Antarctica. *Polar Science, 8*, 342–356. https://doi.org/10.1016/j.polar.2014.07.001.
94. Hildmann, H., & Kovacs, E. (2019). Review: Using unmanned aerial vehicles (UAVs) as mobile sensing platforms (MSPs) for disaster response, civil security and public safety. *Drones, 3*, 1–26. https://doi.org/10.3390/drones3030059.
95. "Inspirational" bear video reveals dark side of filming animals with drones. https://www.nationalgeographic.com/animals/2018/11/drone-brown-bear-video-russia-wildlife-harrassment-news/. Last accessed 2019/11/21.
96. Mulero-Pázmány, M., Jenni-Eiermann, S., Strebel, N., Sattler, T., Negro, J. J., & Tablado, Z. (2017). Unmanned aircraft systems as a new source of disturbance for wildlife: A systematic review. *PLoS One, 12*. https://doi.org/10.1371/journal.pone.0178448.
97. Wild Animal Attacks on drones videos – Drones UAV report. http://dronesuavreport.com/2018/08/04/wild-animal-attacks-on-drones-videos/. Last accessed 2019/12/20.
98. Manfreda, S., McCabe, M. F., Miller, P. E., Lucas, R., Madrigal, V. P., Mallinis, G., Dor, E. B., Helman, D., Estes, L., Ciraolo, G., Müllerová, J., Tauro, F., de Lima, M. I., de Lima, J. L. M. P., Maltese, A., Frances, F., Caylor, K., Kohv, M., Perks, M., Ruiz-Pérez, G., Su, Z., Vico, G., & Toth, B. (2018). On the use of unmanned aerial systems for environmental monitoring. *Remote Sensing, 10*, 1. https://doi.org/10.3390/rs10040641.
99. Tmušić, G., Manfreda, S., Aasen, H., James, M. R., Gonçalves, G., Ben-Dor, E., Brook, A., Polinova, M., Arranz, J. J., Mészáros, J., Zhuang, R., Johansen, K., Malbeteau, Y., de Lima, I. P., Davids, C., Herban, S., & McCabe, M. F. (2020). Current practices in UAS-based environmental monitoring. *Remote Sensing, 12*. https://doi.org/10.3390/rs12061001.
100. Torresan, C., Berton, A., Carotenuto, F., Di Gennaro, S. F., Gioli, B., Matese, A., Miglietta, F., Vagnoli, C., Zaldei, A., & Wallace, L. (2017). Forestry applications of UAVs in Europe: A review. *International Journal of Remote Sensing, 38*, 2427–2447. https://doi.org/10.1080/01431161.2016.1252477.
101. Fraser, B. T., & Congalton, R. G. (2018). Issues in unmanned aerial systems (UAS) data collection of complex forest environments. *Remote Sensing, 10*, 908. https://doi.org/10.3390/rs10060908.
102. McCabe, M. F., Rodell, M., Alsdorf, D. E., Miralles, D. G., Uijlenhoet, R., Wagner, W., Lucieer, A., Houborg, R., Verhoest, N. E. C., Franz, T. E., Gao, H., & Wood, E. F. (2017). The future of earth observation in hydrology. *Hydrology and Earth System Sciences, 21*, 3879–3914. https://doi.org/10.5194/hess-21-3879-2017.
103. Harris, J. M., Nelson, J. A., Rieucau, G., & Broussard, W. P. (2019). Use of drones in fishery science. *Transactions of the American Fisheries Society, 148*, 1–11. https://doi.org/10.1002/tafs.10168.

104. Kelly, J., Kljun, N., Olsson, P. O., Mihai, L., Liljeblad, B., Weslien, P., Klemedtsson, L., & Eklundh, L. (2019). Challenges and best practices for deriving temperature data from an uncalibrated UAV thermal infrared camera. *Remote Sensing, 11*, 567. https://doi.org/10.3390/rs11050567.
105. Aldana-Jague, E., Heckrath, G., Macdonald, A., van Wesemael, B., & Van Oost, K. (2016). UAS-based soil carbon mapping using VIS-NIR (480-1000 nm) multi-spectral imaging: Potential and limitations. *Geoderma, 275*, 55–66. https://doi.org/10.1016/j.geoderma.2016.04.012.
106. INTERACT: Drones Pocket Guide. (2017). *INTERACT drone workshop svalbard.*
107. Storvold, R., Sweatte, C., Ruel, P., Wuennenberg, M., Tarr, K., Raustein, M., Hillesøy, T., Lundgren, T., & Sumich, M. (2015). *Arctic science RPAS operator's handbook.* AMAP.
108. Berkowitz, R. (2014). Drones and the question of the human. *Ethics & International Affairs, 28*, 159. https://doi.org/10.1017/S0892679414000185.
109. Petroleka, M., Sano, Y., Shah, K., Marshall, R., Haines, C., Weber, K., Richards, M., Beckenstein, J., Menonna, F., Thillien, D., Martin, C., Glendinning, T., Richards, E., Grant, J., Shutt, M., Brenden, D., Earnshaw, D., Chia, S., Nix, P., Moss, J., Taylor, R., & Oliva-Velez, D. (2018). *Towards 2050: Megatrends in industry, politics and the global economy 2018 edition.* BMI Research.
110. Cummings, M. L., & Mitchell, P. J. (2008). Predicting controller capacity in supervisory control of multiple UAVs. *IEEE Transactions on Systems, Man, and Cybernetics. Part A Systems Humans., 38*, 451–460. https://doi.org/10.1109/TSMCA.2007.914757.
111. Chen, J. Y. C., Barnes, M. J., & Harper-Sciarini, M. (2011). Supervisory control of multiple robots: Human-performance issues and user-interface design. *IEEE Transactions on Systems, Man, and Cybernetics Part C: Applications and Reviews, 41*, 435–454. https://doi.org/10.1109/TSMCC.2010.2056682.
112. Goodrich, M. A., & Schultz, A. C. (2007). Human-robot interaction: A survey. *Foundations and Trends in Human-Computer Interaction, 1*, 203–275. https://doi.org/10.1561/1100000005.
113. Dierks, T., & Jagannathan, S. (2010). Output feedback control of a quadrotor UAV using neural networks. *IEEE Transactions on Neural Networks, 21*, 50–66. https://doi.org/10.1109/TNN.2009.2034145.
114. Máthé, K., & Buşoniu, L. (2015). Vision and control for UAVs: A survey of general methods and of inexpensive platforms for infrastructure inspection. *Sensors (Switzerland)., 15*, 14887–14916. https://doi.org/10.3390/s150714887.
115. Liu, P., Chen, A. Y., Huang, Y. N., Han, J. Y., Lai, J. S., Kang, S. C., Wu, T. H., Wen, M. C., & Tsai, M. H. (2014). A review of rotorcraft unmanned aerial vehicle (UAV) developments and applications in civil engineering. *Smart Structures and Systems, 13*, 1065–1094. https://doi.org/10.12989/sss.2014.13.6.1065.
116. Lim, Y., Gardi, A., Sabatini, R., Ramasamy, S., Kistan, T., Ezer, N., Vince, J., & Bolia, R. (2018). Cognitive human-machine interfaces and interactions for manned and unmanned aircraft. *Progress in Aerospace Science, 102*, 1–46. https://doi.org/10.1016/j.paerosci.2018.05.002.
117. Zolich, A., Johansen, T. A., Cisek, K., & Klausen, K. (2016). Unmanned aerial system architecture for maritime missions. Design & hardware description. In *2015 workshop on research, education and development of unmanned aerial systems RED-UAS 2015* (pp. 342–350). https://doi.org/10.1109/RED-UAS.2015.7441026.
118. Hedblom, B. (2018). Addressing the low-altitude airspace integration challenge: USS or UTM core? In *ICNS 2018 – Integrated communications, navigation and surveillance conference* (pp. 1–17). https://doi.org/10.1109/ICNSURV.2018.8384933.
119. Johnson, M., Jung, J., Rios, J., Mercer, J., Homola, J., Prevot, T., Mulfinger, D., & Kopardekar, P. (2017). Flight test evaluation of an unmanned aircraft system traffic management (UTM) concept for multiple beyond-visual-line-of-sight operations. In *12th USA/Europe Air Traffic Management Research and Development Seminar.*

120. Kopardekar, P., Rios, J., Prevot, T., Johnson, M., Jung, J., & Robinson, J. E. (2016). Unmanned aircraft system traffic management (UTM) concept of operations. In *16th AIAA aviation technology, integration, and operations conference*. AIAA Aviation.
121. Ancel, E., Capristan, F. M., Foster, J. V., & Condottax, R. C. (2017). Real-time risk assessment framework for unmanned aircraft system (UAS) traffic management (UTM). In *17th AIAA aviation technology, integration, and operations conference, 2017*. AIAA.
122. Battiste, V., Dao, A. Q. V., Strybel, T. Z., Boudreau, A., & Wong, Y. K. (2016). Function allocation strategies for the unmanned aircraft system traffic management (UTM) system, and their impact on skills and training requirements for UTM operators. *IFAC-PapersOnLine., 49*, 42–47. https://doi.org/10.1016/j.ifacol.2016.10.459.
123. Ivancic, W. D., Kerczewski, R. J., Murawski, R. W., Matheou, K., & Downey, A. N. (2019). Flying drones beyond visual line of sight using 4g LTE: Issues and concerns. In *Integrated communications, navigation and surveillance conference 2019-April* (pp. 1–13). IEEE. https://doi.org/10.1109/ICNSURV.2019.8735246.
124. Bloise, N., Primatesta, S., Antonini, R., Fici, G. P., Gaspardone, M., Guglieri, G., & Rizzo, A. (2019). A survey of unmanned aircraft system technologies to enable safe operations in Urban areas. In *2019 international conference on unmanned aircraft systems* (pp. 433–442). https://doi.org/10.1109/icuas.2019.8797859.
125. Beverley, G. (2019). *BVLoS and blockchain – Why distributed ledgers may be the key to unlocking widespread BVLoS operations*. Consortiq.
126. Claesson, A., Fredman, D., Svensson, L., Ringh, M., Hollenberg, J., Nordberg, P., Rosenqvist, M., Djarv, T., Österberg, S., Lennartsson, J., & Ban, Y. (2016). Unmanned aerial vehicles (drones) in out-of-hospital-cardiac-arrest. *Scandinavian Journal of Trauma, Resuscitation and Emergency Medicine, 24*, 1–9. https://doi.org/10.1186/s13049-016-0313-5.
127. Pulver, A., Wei, R., & Mann, C. (2016). Locating AED enabled medical drones to enhance cardiac arrest response times. *Prehospital Emergency Care, 20*, 378–389. https://doi.org/10.3109/10903127.2015.1115932.
128. Claesson, A., Herlitz, J., Svensson, L., Ottosson, L., Bergfeldt, L., Engdahl, J., Ericson, C., Sandén, P., Axelsson, C., & Bremer, A. (2017). Defibrillation before EMS arrival in western Sweden. *The American Journal of Emergency Medicine, 35*, 1043–1048. https://doi.org/10.1016/j.ajem.2017.02.030.
129. Sanfridsson, J., Sparrevik, J., Hollenberg, J., Nordberg, P., Djärv, T., Ringh, M., Svensson, L., Forsberg, S., Nord, A., Andersson-Hagiwara, M., & Claesson, A. (2019). Drone delivery of an automated external defibrillator – A mixed method simulation study of bystander experience. *Scandinavian Journal of Trauma, Resuscitation and Emergency Medicine, 27*, 1–9. https://doi.org/10.1186/s13049-019-0622-6.
130. Clark, D. G., Ford, J. D., & Tabish, T. (2018). What role can unmanned aerial vehicles play in emergency response in the Arctic: A case study from Canada. *PLoS One, 13*, 1–16. https://doi.org/10.1371/journal.pone.0205299.
131. Tierney, M.T. (2016). *Ethics on the fly: Toward a drone-specific code of conduct for law enforcement*. Naval Postgraduate School.
132. Thibbotuwawa, A., Nielsen, P., Bocewicz, G., & Banaszak, Z. (2020). UAVs Fleet Mission planning subject to weather fore-cast and energy consumption constraints. In *Advances in intelligent systems and computing* (pp. 104–114). Springer Verlag. https://doi.org/10.1007/978-3-030-13273-6_11.
133. Schenkelberg, F. (2016). How reliable does a delivery drone have to be? In *Proceedings – Annual reliability and maintainability symposium*. Institute of Electrical and Electronics Engineers Inc. https://doi.org/10.1109/RAMS.2016.7448054.
134. Nex, F., & Remondino, F. (2014). UAV for 3D mapping applications: A review. *Applied Geomatics, 6*, 1–15. https://doi.org/10.1007/s12518-013-0120-x.
135. Al-Kaff, A., Martín, D., García, F., de la Escalera, A., & María Armingol, J. (2018). Survey of computer vision algorithms and applications for unmanned aerial vehicles. *Expert Systems with Applications, 92*, 447–463. https://doi.org/10.1016/j.eswa.2017.09.033.

136. Liu, Z., Zhang, Y., Yu, X., & Yuan, C. (2016). Unmanned surface vehicles: An overview of developments and challenges. *Annual Reviews in Control, 41*, 71. https://doi.org/10.1016/j.arcontrol.2016.04.018.
137. Imdoukh, A., Shaker, A., Al-Toukhy, A., Kablaoui, D., & El-Abd, M. (2017). Semi-autonomous indoor firefighting UAV. In *2017 18th international conference on advanced robotics*. ICAR. https://doi.org/10.1109/ICAR.2017.8023625.
138. Padró, J. C., Muñoz, F. J., Planas, J., & Pons, X. (2019). Comparison of four UAV geo-referencing methods for environmental monitoring purposes focusing on the combined use with airborne and satellite remote sensing platforms. *International Journal of Applied Earth Observation and Geoinformation, 75*, 130–140. https://doi.org/10.1016/j.jag.2018.10.018.
139. Cho, Y. (2014). Lost in debate: The safety of domestic unmanned aircraft systems. *Journal of Strategic Security, 7*, 38–56.
140. Sullivan-Nightengale, D. (2015). Unmanned aerial systems: Risks & opportunities in the workplace. *Professional Safety, 60*, 34–42.
141. Sanjab, A., Saad, W., & Başar, T. (2017). Prospect theory for enhanced cyber-physical security of drone delivery systems: A network interdiction game. *arXiv,* 0–5.
142. Altawy, R., & Youssef, A. M. (2017). Security, privacy, and safety aspects of civilian drones: A survey. *ACM Transactions on Cyber-Physical Systems, 1(2)*, 1–25. https://doi.org/10.1145/3001836.
143. Hamilton, B.A. (2020). *Updated with additional testing and analysis executive summary*. Booz Allen for PrecisionHawk.
144. Mozaffari, M., Saad, W., Bennis, M., & Debbah, M. (2017). Wireless communication using unmanned aerial vehicles (UAVs): Optimal transport theory for hover time optimization. *arXiv, 16*, 8052–8066.
145. Kramar, V. (2020). UAS (drone) in Response to Coronavirus. In *27th conference of open innovation association, FRUCT* (pp. 90–100). Trento, Italy: IEEE Computer Society. https://doi.org/10.23919/FRUCT49677.2020.9211075.
146. European Union Aviation Safety Agency: Easy access rules for unmanned aircraft systems (Regulations (EU) 2019/947 and (EU) 2019/945) (2021) (pp. 1–292).
147. Hodgson, J. C., & Koh, L. P. (2016). Best practice for minimising unmanned aerial vehicle disturbance to wildlife in biological field research. *Current Biology, 26*, R404–R405. https://doi.org/10.1016/j.cub.2016.04.001.
148. ASTM. (2018). *New practice for general operations manual for professional operator of light Unmanned Aircraft Systems (UAS)*. ASTM International.
149. JARUS. (2019). *JARUS guidelines on Specific Operations Risk Assessment (SORA)*. Joint Authorities for Rulemaking of Unmanned Systems.
150. Arctic Drone Labs – Finnish Drone Expertise. https://www.arcticdronelabs.com/. Last accessed 2020/06/05.

Applying an Icing Wind Tunnel for Drone Propeller Research, Validation of New Measurement Instrument

Petri Suurnäkki, Tuomas Jokela, and Mikko Tiihonen

Abstract Unmanned aerial vehicles have increased in popularity in recent years, especially the numbers of small multicopters. At the same time, icing research of such systems has been left behind, and results especially for propellers in this scale and in VTOL configuration are few. While the numbers of such systems have grown, also their usage in cold and icing conditions has increased.

A lot of research has been conducted for full-size airplanes and rotorcraft, but for drones the Reynolds numbers are relatively low in comparison. For this reason, it is important to research these systems in order to develop anti-icing methods and operate drones safely in all weather conditions. Also, currently used numerical tools are developed and validated for high Reynolds number conditions, but such validation has not yet been conducted for low Reynolds number flows.

VTT has operated an icing wind tunnel since 2009 primarily for experimental research in wind power technologies. Part of this line of research, methods for preventing icing of wind turbine blades, has been developed, and numerical tools developed in-house have been validated.

For developing the icing wind tunnel capabilities, a propeller dynamometer was added as a research instrument. This provides the means to research propellers used in drones to be researched in the wind tunnel. During the commissioning of the instrument, experiments in warm and dry conditions were conducted for validation and repeatability purposes. Experiments showed that the thrust measurements were accurate and repeatable, but torque measurement requires more development.

Keywords Icing Wind Tunnel · Propeller Dynamometer · Wind Tunnel Instrumentation · Aerodynamic Icing · Wind Tunnel Testing

P. Suurnäkki (✉)
Aalto University, Espoo, Finland
e-mail: petri.suurnakki@aalto.fi

T. Jokela · M. Tiihonen
VTT Technical Research Centre of Finland, Espoo, Finland
e-mail: tuomas.jokela@vtt.fi; mikko.tiihonen@vtt.fi

T. Lipping et al. (eds.), *New Developments and Environmental Applications of Drones*,
https://doi.org/10.1007/978-3-030-77860-6_2

1 Background

Unmanned aerial vehicles (UAVs) or drones, and especially those with vertical take-off and landing (VTOL) capabilities, have increased in usage rapidly in the recent years. As with larger aircraft, icing is also a serious hazard to drones of various sizes and architectures. UAVs have become a valuable tool for various applications ranging from remote sensing duties to transportation of goods. Thus, ensuring their operational safety in these kinds of conditions, it is essential to avoid unnecessary hazards in the case they have to be operated during icing threat. Also, in order to design and operate UAVs in various conditions, knowledge on the low Reynolds number propeller performance and especially the change in performance due to icing is essential. Today, there is limited research on UAV and low Reynolds number propeller icing, and no mature icing protection methods for these exist. For these reasons, research conducted especially on low Reynolds number propeller performance in icing conditions needs to be performed. Hence, fitting an icing wind tunnel with proper instrumentation for propeller research is important.

Icing is a common threat in aviation and poses a threat for large aircraft and rotorcraft, as it does to smaller drones. Hence, throughout the years, a lot of research has been carried out for general aviation, and a good understanding of ice formation and its effects on aerodynamics for fixed-wing aircraft and rotorcraft have been established [1–4]. Similar large-scale research has also been extensively conducted on wind turbine rotors, where in-depth knowledge of aerodynamic icing also exists [5–7].

Testing is also important for validation of numerical tools, as numerical tools can provide key analysis more affordably than fully experimental workflow. Larger numerical model validations have been conducted previously by instances such as AGARD [8], for the scale of large commercial aircraft. Meaning that while they have been validated thoroughly and proven to work, they have been developed for high Reynolds number. While there is some research conducted on the numerical tools of low Reynolds number icing [9, 10], the amount of research and validation of these tools is still in short supply, and more experimental research is still required in order to properly validate these tools.

For smaller drones, very few icing protection systems exist, and for development of these, experimental work is crucial. On the contrary, large-scale aircraft have mature icing protection systems due to the large volume of research conducted [11]. Some development for fixed-wing UAV anti-icing solutions has already been done [12–17], but the amount of research results in this scale is still limited. Additionally, some research has been conducted for propellers of this scale operating in icing conditions [18–22].

1.1 Aerodynamic Icing

Aerodynamic icing happens when supercooled water droplets impinge and freeze on unprotected surfaces [1]. Physics of ice accretion can be divided into three topics: droplet impingement, ice accretion, and ice adhesion [23], but often it is convenient to consider the first two distinct parts [1].

For droplet impingement, the rate at which droplets strike the aircraft surfaces is the product of several variables [1]. The water catch efficiency is dependent on the liquid water content (LWC) of the cloud, size and shape of the body, and airspeed [1]. Additionally, there is limited effect from the ambient pressure due to altitude and temperature for the collection efficiency [1].

Ice accretion is the rate at which the striking water droplets freeze to form ice on the unprotected surface, which is primarily governed by heat transfer from the surface of the body [1]. Two main ice types forming are rime and glaze ice, but also certain types of mixed ice of these two can be encountered [23]. Additionally, at warm temperatures or high speed, slushy ridge of ice called beak ice forms, and this is the third main type of ice accretion that may form on wings or rotor blades [1]. These main ice types are presented in Fig. 1.

If aerodynamic surfaces are not protected, icing causes aerodynamic penalties by increasing drag and reducing the lift of lifting surfaces [23]. The changes of aerodynamic effects can be separated to surface roughness and changes of the shape of the body due to icing [23]. While the lift reduces and drag increases, the aircraft's weight also grows, and combination of these may result in the inability to continue flying [24, 25]. For rotorcraft, the icing affects the handling due to the same reasons [1]. The icing also affects the pitching moment characteristics, which may also increase the control loads [1].

1.2 Wind Tunnels

Experimentation is an important aspect of aerospace engineering, and wind tunnels are one of the options. Low-speed wind tunnels are possible to use to test models prepared early in design cycles, offering also the full complexity of real fluid flow [26]. Wind tunnels provide large amounts of reliable data and are a rapid and accurate mean to conduct aerodynamic research [26]. In addition to numerical tools, wind tunnels are an essential tool in engineering, both for model testing of new designs and basic aerodynamic research [27, 28]. Wind tunnel testing has also grown interest in other branches of industry and science when it comes to low-speed aerodynamics [28].

Wind tunnels are usually defined by three main criteria: maximum wind speed, flow uniformity, and turbulence intensity [28]. For the structure of wind tunnels, there are two basic types of wind tunnels: the open-circuit wind tunnel and the closed-circuit wind tunnel [26]. Additionally, there are also two configurations of

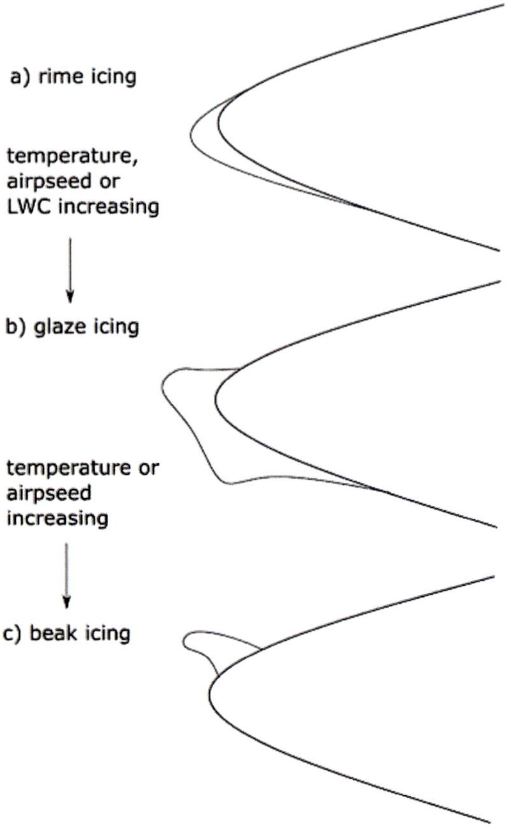

Fig. 1 Illustration of the main ice types. Redrawn from [1]

basic test sections: open or closed [26]. The measurement equipment also varies in wind tunnels varying from observing tools to balances and acoustic measurement tools [26, 28]. Force measurements are done with force balances, which can be categorized into internal and external balances [28]. Thus, wind tunnels vary a lot, and generally, the type of wind tunnel and instrumentation depend on the purpose of the tunnel and available funds [26].

While there are many specialized wind tunnels, one of the types is the icing wind tunnel. These are used to research the effects of aerodynamic icing and impact on aircraft performance. Icing wind tunnels have a refrigeration system to decrease the air temperature and water atomizers upstream of the test section for water droplet production [26]. Both open-circuit and closed-circuit icing wind tunnels exist, and one example of the closed-circuit icing wind tunnels is the NASA 6 x 9-ft Icing Research Tunnel at Lewis Research Center [26]. This is a closed-loop low-speed wind tunnel, where the shell of the tunnel is heavily insulated to keep circulating air in the tunnel cold [26].

2 Applying the VTT Icing Wind Tunnel for Drone Research

VTT Icing Wind Tunnel (IWT) is an open-return type subsonic wind tunnel located in a climate-controlled room. The IWT has a 3×3 water spraying nozzle matrix located in the non-contracted part of the wind tunnel. With the added contraction, the contraction ratio is 3.19:1, but the wind tunnel can also be operated as an open jet facility. Contracted test section has a nominal size of 700×700 mm in cross-section and 1000 mm in length. The uncontracted cross-section is 1250×1250 mm. A schematic of the wind tunnel is presented in Fig. 2.

Wind tunnel can achieve wind speeds up to 50 m/s, but only for limited periods, whereas wind speeds up to 20 m/s can be achieved continuously. It is possible to operate the facility between −25 and 30 °C. LWC range is between 0.1 and 1.0 g/m^3 with droplet sizes (MVD) between 12 and 30 μm. Turbulence intensity (TI) in the test section varies between 0.6 and 1.3%. The wind tunnel is capable of producing all three ice types: rime ice, glaze ice, and mixed ice.

Wind speeds are calibrated with a cup anemometer, and LWC calibrations are conducted using ISO 12494 procedure [29]. LWC and MVD can be changed by varying the water flow and atomizing air flow to the nozzle matrix.

In order to apply the wind tunnel better for drone research first, the existing capabilities were analyzed in order to create requirements for new capabilities. Based on this analysis, a new measurement device, a propeller dynamometer, was fitted into the wind tunnel. In this chapter, the dynamometer was validated against the existing results, and repeatability of the measurements in cold conditions was investigated.

2.1 Existing Capabilities

As the IWT has been in use for over a decade, it already has the existing capabilities regarding drone research. Obviously, the wind tunnel is possible to be used to study ice accretion as the wind tunnel conditions are well known. For example, there is

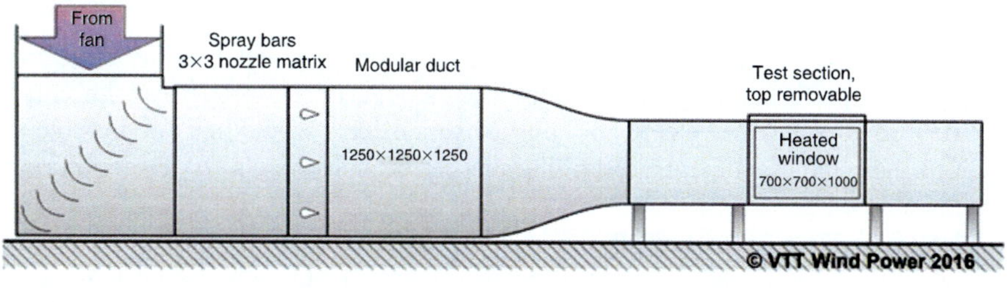

Fig. 2 VTT icing wind tunnel schematic

a set of well-verified standard conditions, and the droplet size is well known in a variety of conditions.

Due to the earlier research in wind power domain, it is possible to do experiments with airfoil sections and sensors in the tunnel. Also, both of these in well-lit conditions provide means for good imaging. There is also a possibility to use a single-axis balance for force measurements. For the size of the tunnel, it is possible to test full-drone systems for ice accretion and even possible to fly drones up to 2 kg maximum take-off weight (MTOW) with the open jet.

A complete list of the existing capabilities of the wind tunnel:

- Ice accretion, well-verified standard conditions
- Good imaging possibilities of the experiments
- Experiments for airfoil sections
- Experiments for sensors
- Possible to test full-drone systems
  - Maximum 2 kg MTOW systems possible to fly with the open jet
- Single-axis force measurement

2.2 *Requirements for New Capabilities*

While the wind tunnel already had capabilities suitable for drone research in icing conditions, it was unable to be used for propeller research. Especially with VTOL drones, propeller performance is crucial for flying safely in cold climate. Similarly, propeller performance also limits fixed-wing drones flight envelopes in icing conditions. Thus, it was deemed crucial to be able to research propeller performance in the icing wind tunnel, and requirements for new capabilities aspire from this.

Basic requirement was to be able to test propeller performance. As propeller performance is characterized by the thrust coefficient and power coefficient, it was thus required to be able to measure the thrust and torque of the propeller. Additionally, it was good to be able to monitor the power consumption of the system in addition to the torque. Also, since icing can create imbalance into the propeller, it was required to be able to measure the vibrations of the propeller.

As the icing is a dynamic process and thus in order to monitor the development of propeller performance, it was required to be able to test propeller with either constant rotational speed (RPM) or constant thrust. This would make it possible to monitor how the coefficients describing the performance begin to develop during the icing process.

Additionally, in order to keep the costs low regarding the new capabilities and operating the wind tunnel, it was required that the new solution is also cost efficient.

List of requirements for the new testing capabilities:

- Testing propeller performance
  - Thrust measurement
  - Torque measurement
  - Vibration measurement
  - Power consumption
- Possible to test propellers with constant RPM or constant thrust
- Cost efficient solution

2.3 Propeller Dynamometer

To improve the drone research capabilities based on the requirements, a propeller dynamometer was fitted as a new measurement instrument. The instrument was a commercial Series 1585 propeller dynamometer from RCbenchmark. This commercial product got an in-house modification. Main part of the modification was adding a protective shroud to cover the electronics, load cells, and other sensors from ice accretion. The "VTT mod" propeller dynamometer was then fitted into the contracted test section of the wind tunnel as seen in Fig. 3.

The dynamometer comes with a control board, which works as the data acquisition system and controller, but it requires additional electronics to function. The complete system consisted of DualSky XC-65-Lite electronic speed controller (ESC) and a DualSky XM5010TE-4 motor. The whole system was powered by a TDK-Lambda GEN60-55-3P400 power supply, providing constant 11.1 V for the system.

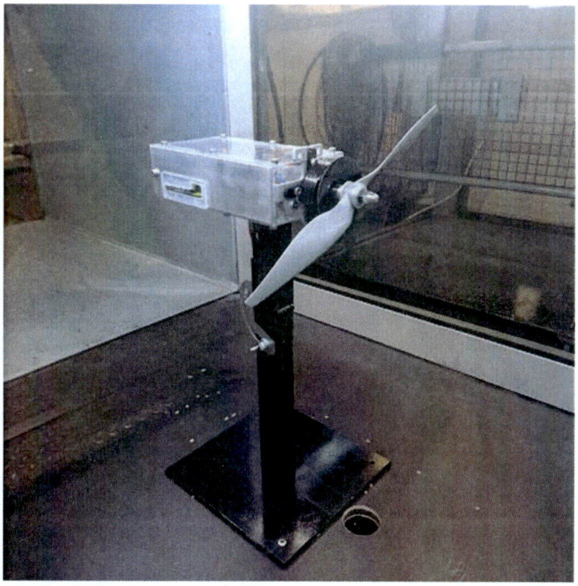

Fig. 3 "VTT mod" propeller dynamometer fitted into the wind tunnel test section

With this new instrument, it is possible to measure the thrust of the propeller and the reactional torque. Additionally, electrical power, through voltage and current, can be recorded. The dynamometer also has accelerometers for measuring vibrations. Propellers between 6" and 12" can be tested with this setup in the size limitations posed by the wind tunnel test section.

With the included graphical user interface (GUI), it is also possible to create scripts for the system. In-house script for a PID-controller was created in order to run experiments with constant RPM or thrust.

3 Experimental Setup

In order to test the accuracy and performance of the new instrument, validation tests in warm and dry conditions were conducted. In addition to this, the repeatability of the system was tested in icing conditions. For these experiments, a well-characterized propeller was chosen as the test specimen.

3.1 APC Propeller

For validating the new measurement device, APC Thin Electric 11×8 propeller was chosen, as this propeller has been researched earlier by Brand and Selig [30] providing good baseline data and it has good availability. For this validation study, 3 propeller specimens of the same type were used, and they were used in rotation during individual runs. Figure 4 represents the geometric characteristics of chord c, sweep s as the ratio of the radius R and twist β provided by the propeller manufacturer. All of these were represented as a function of the ratio of the local radius r and propeller radius. The root airfoil is Eppler E63 transitioning to NACA 4412 at the tip.

3.2 Procedures for Validation and Icing Conditions Repeatability Tests

The validation tests were run in warm and dry conditions, where the propeller was running at a constant RPM and wind speed was varied for different advance ratios. For these experiments, the RPMs used were 4000 and 6000, which would yield Reynolds numbers between 55,000 and 95,000. The temperature of the facility was kept at 20 ° C. Data was collected for 5 min for each discrete step of advance ratio.

In order to test the repeatability of the measurement instrument in cold and wet conditions, a light icing condition was chosen. For this, the facility temperature was

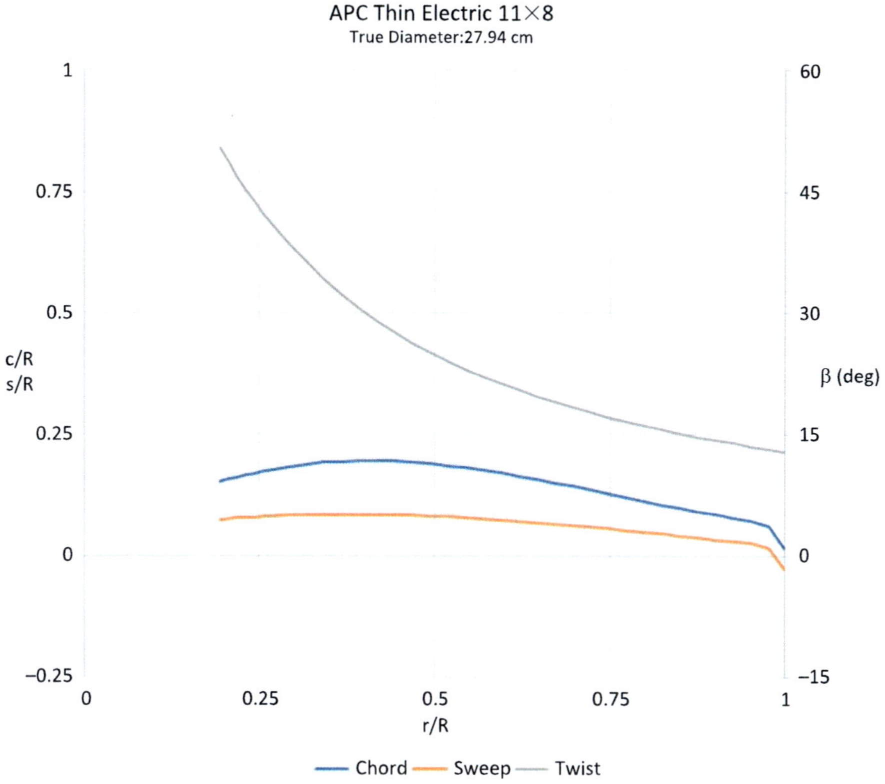

Fig. 4 Geometric characteristics of the APC thin electric 11×8 propeller. Plotted from manufacturer provided data

lowered to −5 °C and air humidity increased to the range of 85–100%. The liquid water content LWC was chosen to be 0.1 g/m^3, which is a fairly light condition for icing. The LWC was validated with the ISO 12,494 [29] procedure. Similarly to the validation tests in warm conditions, the propeller was kept rotating at a constant RPM of 4000 and 6000, and wind speed varied for different advance ratios. The test was run for 5 min or until the first shedding event.

3.3 Data Reduction

The majority of the data reduction was already handled by the dynamometer control board, by converting the electrical measurements into physical units. Thus, the system already provides measurements for thrust T, torque Q, rotational speed n, system voltage U, and current I. The propeller performance data is derived from the measured quantities by non-dimensionalizing them for advance ratio J, thrust coefficient C_T, and power coefficient C_P. These derivations utilize the air density ρ, rotational speed n, wind speed V, and propeller diameter D. As all cases were run non-static, the coefficient is plotted against the advance ratio and time. These non-dimensional quantities are defined as

$$J = \frac{V}{nD} \tag{1}$$

$$C_T = \frac{T}{\rho n^2 D^4} \tag{2}$$

$$C_P = \frac{P}{\rho n^3 D^5}. \tag{3}$$

The power coefficient is derived from the power P, and most commonly, this is calculated from the mechanical power, but also the electrical power can be used. Mechanical and electrical powers are given by

$$P = 2\pi n Q \tag{4}$$

$$P = UI. \tag{5}$$

4 Results and Discussion

Generally, the thrust measurement works accurately, but during the validation tests, it was noticed that the reactional torque measurement did not provide accurate and consistent results. As the system measures reactional torque with two load cells, the structure may pose variation in the measurements. Especially, the hinges that connect the motor mount to the load cells could be improved in quality as they were noticed to cause unpredictable hysteresis. This is most likely the main cause of inconsistent torque results. Thus, more development on the torque measurement needs to be done, while the thrust measurement was validated. Development may be done by fitting a torque transducer or developing the structure of the dynamometer better. Results from the electrical power measurements were observed to be repeatable, but they are not comparable against the baseline results from Brandt and Selig [30].

There are some general observations to be pointed out from the repeatability tests in icing conditions. Generally, the thrust measurements in icing conditions are repeatable at low advance ratios throughout the test run. At higher RPM, initially the measurements begin in repeatable conditions, but due to the dynamic process of ice accretion there is some variation in the results once higher amounts of ice have accreted.

4.1 Validation Results

As mentioned above, the thrust measurements at reference conditions are repeatable and match the reference baseline results by Brandt and Selig [30]. These thrust measurement validation results are presented in Figs. 5 and 6.

Observing these results validates that the thrust measurement is accurate and repeatable. They match well to the baseline data provided for the propeller. For these figures and each step of advance ratio, the test was run 5 times for the earlier mentioned 5 min. These results were then averaged out for the final figures. Between the runs, there was also only minor variation to be observed.

4.2 Repeatability Results in Icing Conditions

Figures 7, 8, 9, 10, 11, 12, 13, and 14 present the results of the repeatability tests in icing conditions. Thrust measurements at 4000 RPM with two advance ratios are presented in Figs. 7 and 8, and representative electrical power measurements are in Figs. 11 and 12. Similarly, Figs. 9 and 10 show the thrust measurements at 6000 RPM for two advance ratios, complemented by electrical power measurements

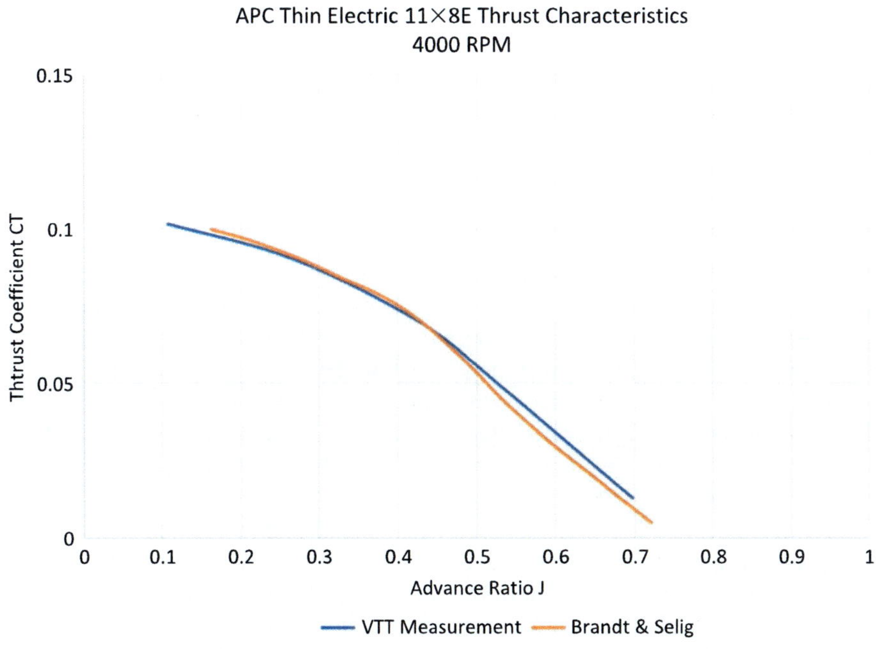

Fig. 5 Thrust characteristics at 4000 RPM. Average from several measurements

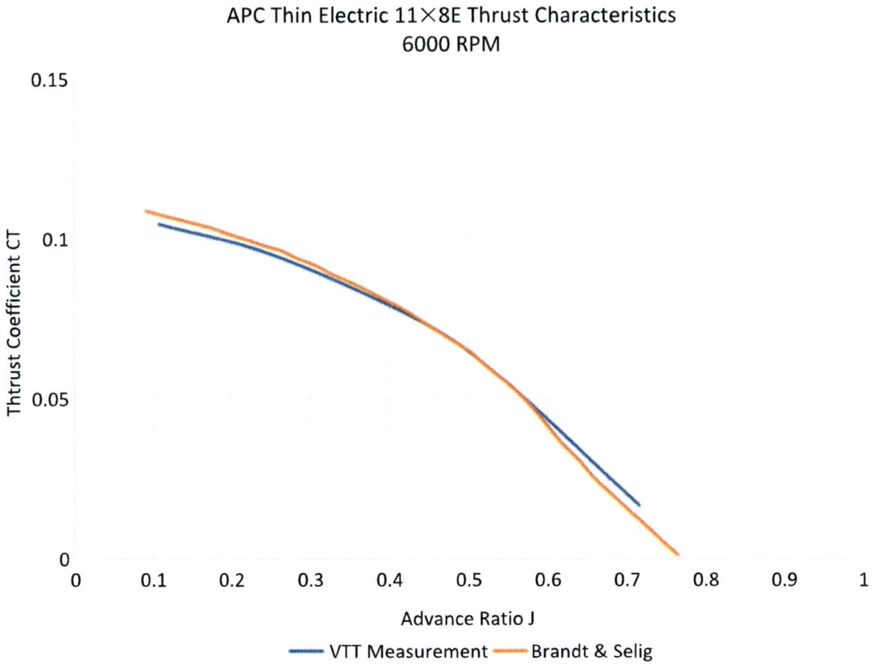

Fig. 6 Thrust characteristics at 6000 RPM. Average from several measurements

in Figs. 13 and 14. As the reactional torque measurement was observed to require more development in the validation tests, the electrical power measurements were used instead.

Observing Figs. 7, 8, 9, and 10 shows that the thrust measurements repeat well also in icing conditions. Especially at the lower RPM throughout the measurement period, but with the higher RPM due to the icing process variation at the later stages can be observed. Still, in the beginning for both RPMs, the measurements are repeatable. Ice shedding can be observed in Fig. 10 when the measurement has ended before 300 s. Some ice shedding can also be observed in Fig. 9 in a single run when thrust measurement became more noisy. In the scope of this chapter, further observations about ice shedding or drop of thrust coefficients are not done. For the scope of developing the wind tunnel capabilities, repeatability of the tests is considered at this point, but analyzing what the results would mean in the scope of drone operations is not relevant.

From Figs. 11, 12, 13, and 14, it can also be observed that the electrical thrust measurements are repeatable. For both RPMs, this trend is observable. In Fig. 13, the ice shedding can also be observed from some of the runs as increased measurement amplitude. But regardless, the electrical power measurements especially at the beginning are well repeatable, and variation is shown due to the dynamic process of icing.

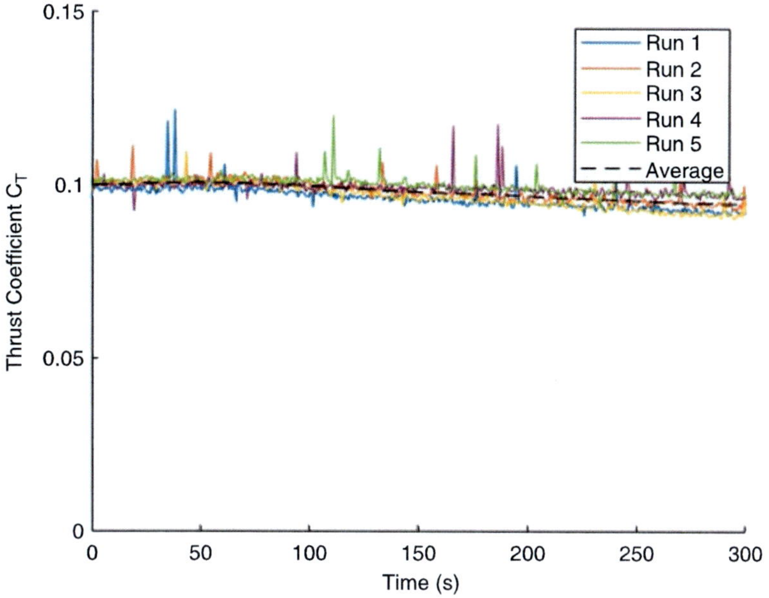

Fig. 7 Thrust measurements at 4000 RPM with advance ratio 0.11

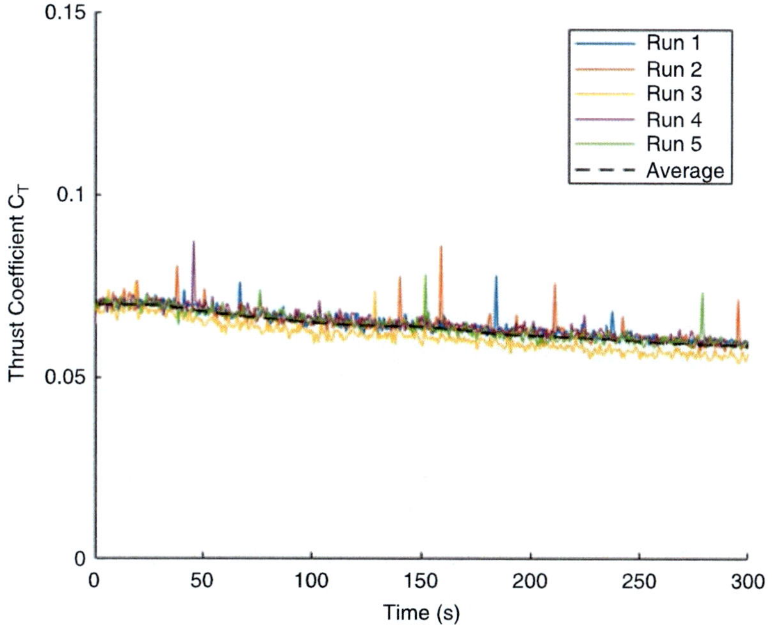

Fig. 8 Thrust measurements at 4000 RPM with advance ratio 0.43

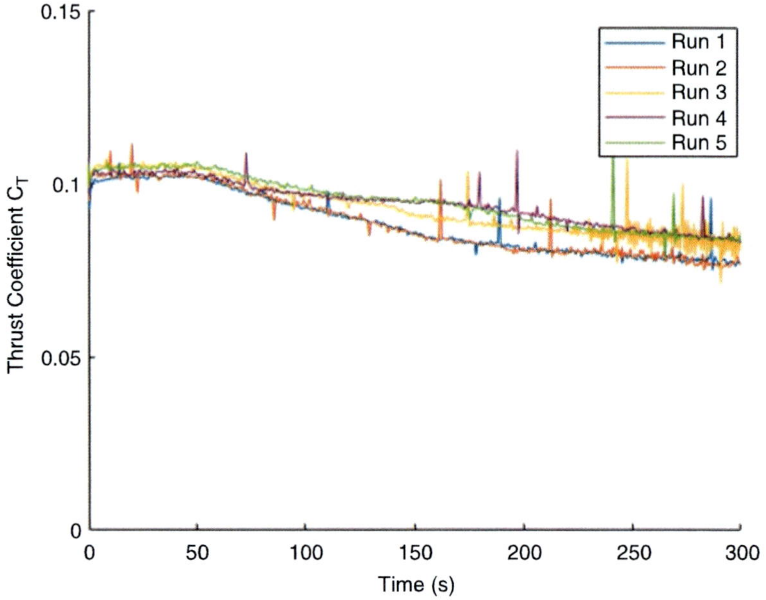

Fig. 9 Thrust measurements at 6000 RPM with advance ratio 0.11

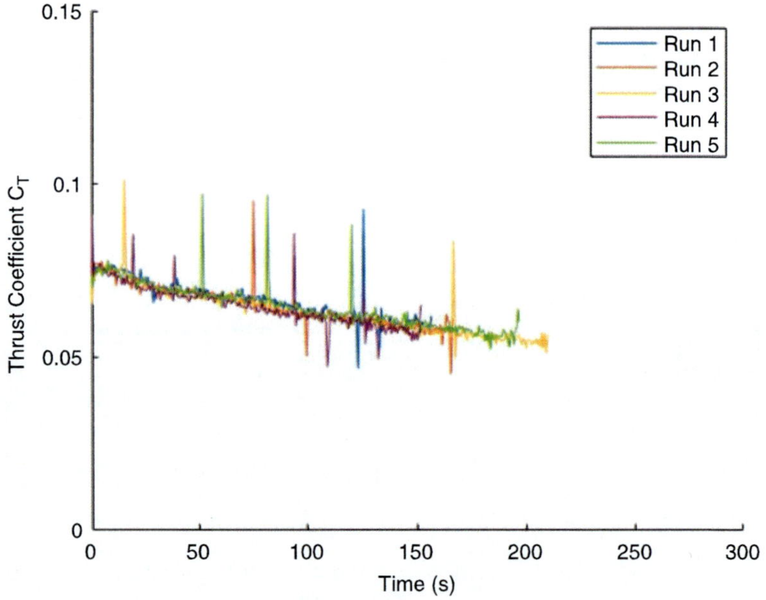

Fig. 10 Thrust measurements at 6000 RPM with advance ratio 0.43

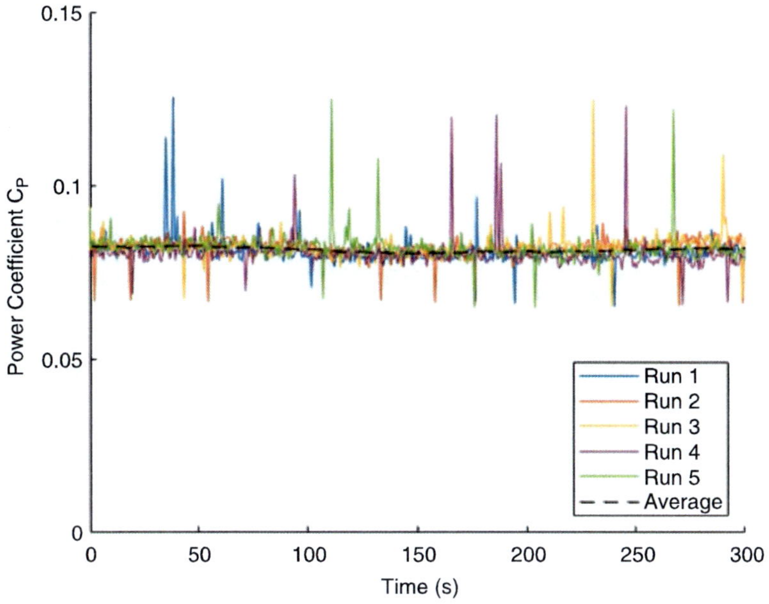

Fig. 11 Electrical power measurements at 4000 RPM with advance ratio 0.11

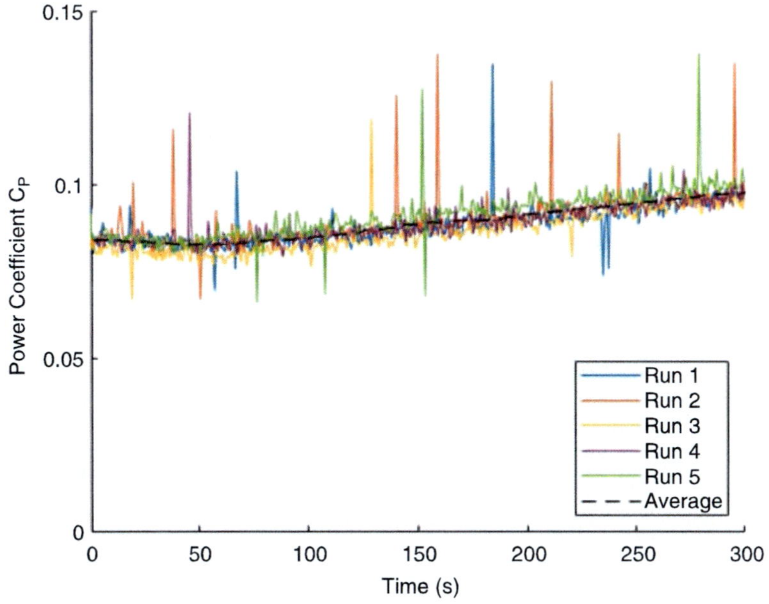

Fig. 12 Electrical power measurements at 4000 RPM with advance ratio 0.43

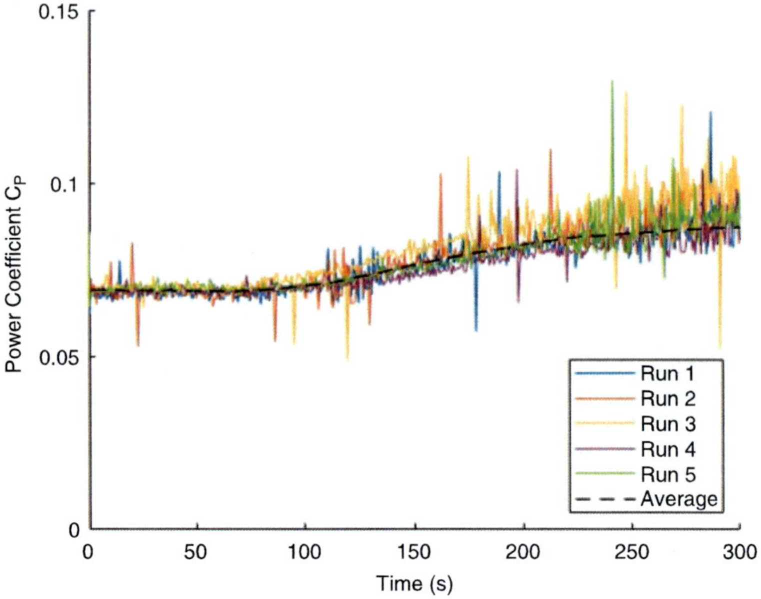

Fig. 13 Electrical power measurements at 6000 RPM with advance ratio 0.11

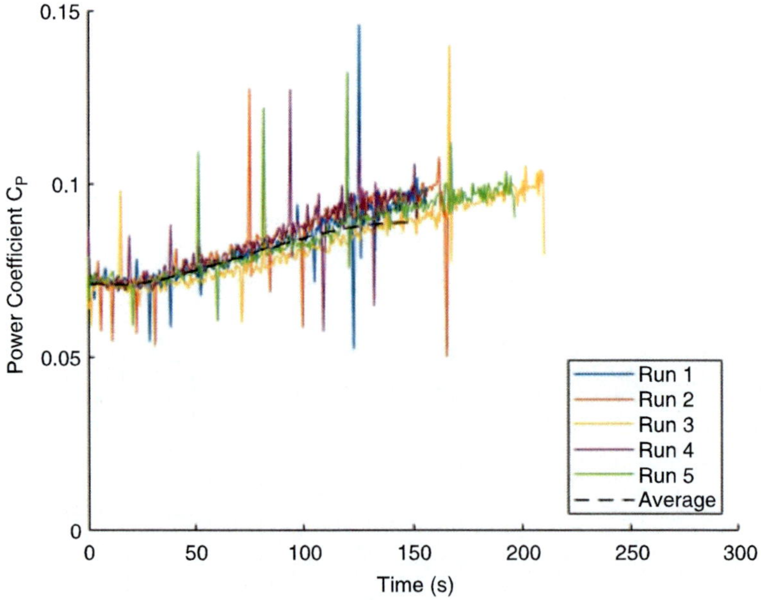

Fig. 14 Electrical power measurements at 6000 RPM with advance ratio 0.43

5 Conclusion

Fitting an icing wind tunnel with a propeller dynamometer for improving drone research capabilities has been presented. While the wind tunnel already had the existing capabilities suitable for drone research, the instrumentation presented provides the means to conduct research on low-Reynolds number propeller performance.

Commercial products provide cost-efficient solutions, but modifications to them are to be expected. They also offer good user interfaces and ease of use but may require further development to completely suit the purpose. The balance of resource usage between modified commercial solution and completely custom solution is to be looked.

It is also important to validate the new instrument against the existing data and research the repeatability of the results. For this purpose, the new instrument was fitted with a propeller that had good existing baseline data available. With the validation tests, it was found that the thrust measurement is accurate and repeatable, but the reactional torque measurement requires further development. In tests performed in cold conditions, the thrust measurements were repeatable, and same observation was done for the electrical power measurements.

Developing the wind tunnel capabilities with the new instruments paves the road for further research of drones in icing environments. Thus, research is continued by exploring drone propellers performance in various icing conditions. Additionally, it is possible to develop methods to protect the propellers from icing to be able to mitigate the safety risk and extend the flight envelope in icing conditions.

References

1. Gent, R., Dar, N., & Cansdale, J. (2000). Aircraft icing. *Philosophical Transactions of the Royal Society of London. Series A: Mathematical, Physical and Engineering Sciences, 358*(1776), 2873–2911. https://doi.org/10.1098/rsta.2000.0689
2. Cebeci, T., & Kafyeke, F. (2003). Aircraft icing. *Annual Review of Fluid Mechanics, 35*(1), 11–21. https://doi.org/10.1146/annurev.fluid.35.101101.161217
3. Miller, T. L., & Bond, T. H. (1989). Icing research tunnel test of a model helicopter rotor. In *45th Annual Forum and Technology Display*.
4. Shaw, R., & Richter, G. (1985). The UH-1H helicopter icing flight test program - An overview. In *23rd Aerospace Sciences Meeting* (p. 338). https://doi.org/10.2514/6.1985-338
5. Jasinski, W. J., Noe, S. C., Selig, M. S., & Bragg, M. B. (1998). Wind turbine performance under icing conditions. In *AIAA, 35th Aerospace Sciences Meeting & Exhibit*. https://doi.org/10.1115/1.2888048
6. Makkonen, L., Laakso, T., Marjaniemi, M., & Finstad, K. J. (2001). Modelling and prevention of ice accretion on wind turbines. *Wind Engineering, 25* (1), 3–21. https://doi.org/10.1260/0309524011495791
7. Homola, M. C., Wallenius, T., Makkonen, L., Nicklasson, P. J., & Sundsbø, P. A. (2010). The relationship between chord length and rime icing on wind turbines. *Wind Energy, 13*(7), 627–632. https://doi.org/10.1002/we.383

8. 20, F. D. P. W. G. (1997). *Ice Accretion Simulation AGARD Advisory Report 344*. Advisory Group for Aerospace Research and Development (AGARD), NATO.
9. Hann, R. (2018). UAV icing: Comparison of LEWICE and FENSAP-ICE for ice accretion and performance degradation. In *Atmospheric and Space Environments Conference* (p. 2861). https://doi.org/10.2514/6.2018-2861
10. Hann, R. (2019). UAV icing: Comparison of LEWICE and FENSAP-ICE for anti-icing loads. In *AIAA Scitech 2019 Forum* (p. 1286). https://doi.org/10.2514/6.2019-1286
11. Goraj, Z. (2004). An overview of the deicing and anti-icing technologies with prospects for the future. In *24th International Congress of Aeronautical Sciences*, vol. 29.
12. Sørensen, K. L. (2016). *Autonomous Icing Protection Solution for Small Unmanned Aircraft: An Icing Detection, Anti-Icing and De-Icing Solution*, Ph.D. Thesis.
13. Hann, R., Wenz, A., Gryte, K., & Johansen, T. A. (2017). Impact of atmospheric icing on UAV aerodynamic performance. In *Workshop on Research, Education and Development of Unmanned Aerial Systems (RED-UAS)* (pp. 66–71). Piscataway: IEEE. https://doi.org/10.1109/RED-UAS.2017.8101645
14. Hann, R. (2019). *UAV Icing: Ice Accretion Experiments and Validation*. Tech. Rep., SAE Technical Paper. https://doi.org/10.4271/2019-01-2037
15. Hann, R., Borup, K., Zolich, A., Sorensen, K., Vestad, H., Steinert, M., & Johansen, T. (2019). *Experimental Investigations of an Icing Protection System for UAVs*. Tech. Rep., SAE Technical Paper. https://doi.org/10.4271/2019-01-2038
16. Hann, R., Hearst, R. J., Sætran, L. R., & Bracchi, T. (2020). Experimental and numerical icing penalties of an S826 airfoil at low reynolds numbers. *Aerospace, 7*(4), 46. https://doi.org/10.3390/aerospace7040046
17. Broeren, A. P., Woodard, B., Diebold, J. M., & Moens, F. (2017). Low-Reynolds number aerodynamics of an 8.9% scale semispan swept wing for assessment of icing effects. In *9th AIAA Atmospheric and Space Environments Conference* (p. 4372). https://doi.org/10.2514/6.2017-4372
18. Liu, Y., Li, L., & Hu, H. (2017). An experimental study on the transient heat transfer and dynamic ice accretion process over a rotating UAS propeller. In *9th AIAA Atmospheric and Space Environments Conference* (p. 4474). https://doi.org/10.2514/6.2017-4474
19. Liu, Y., Li, L., & Hu, H. (2018). Effects of ice accretion on the aerodynamic performance and wake characteristics of an UAS propeller model. In *Atmospheric and Space Environments Conference* (p. 3496). https://doi.org/10.2514/6.2018-3496
20. Liu, Y., Li, L., Ning, Z., Tian, W., & Hu, H. (2018). Experimental investigation on the dynamic icing process over a rotating propeller model. *Journal of Propulsion and Power, 34*(4), 933–946. https://doi.org/10.2514/1.B36748
21. Liu, Y., Li, L., Li, H., & Hu, H. (2018). An experimental study of surface wettability effects on dynamic ice accretion process over an UAS propeller model. *Aerospace Science and Technology, 83*, 164–172. https://doi.org/10.1016/j.ast.2017.12.003
22. Liu, Y., Li, L., Chen, W., Tian, W., & Hu, H. (2019). An experimental study on the aerodynamic performance degradation of a UAS propeller model induced by ice accretion process. *Experimental Thermal and Fluid Science, 102*, 101–112. https://doi.org/10.1016/j.expthermflusci.2018.11.008
23. Heinrich, A., Ross, R., Zumwalt, G., Provorse, J., & Padmanabhan, V. (1991). *Aircraft Icing Handbook*, vol. 1. Gates Learjet Corp Wichita KS.
24. Hoerner, S. F. (1965). Fluid-dynamic drag: Theoretical. *Experimental and Statistical Information.*
25. Hoerner, S. F. (1976). Fluid-dynamic lift: Practical Information on Aerodynamic and Hydrodynamic Lift. *L.A. Hoerner*
26. Barlow, J. B., Rae, W. H., & Pope, A. (1999). *Low-speed wind tunnel testing*. Hoboken: Wiley.
27. Mehta, R. D., & Bradshaw, P. (1979). Design rules for small low speed wind tunnels. *Aeronautical Journal, 83*(827), 443–449.

28. Hernández, M. A. G., López, A. I. M., Jarzabek, A. A., Perales, J. M. P., Wu, Y., & Xiaoxiao, S. (2013). Design methodology for a quick and low-cost wind tunnel. *Wind Tunnel Designs and Their Diverse Engineering Applications, 1*, 3–26. http://doi.org/10.5772/54169
29. ISO 12494:2001(E). (2001). Atmospheric icing of structures, Standard. International Organization for Standardization, Geneva, CH.
30. Brandt, J., & Selig, M. (2011) Propeller performance data at low Reynolds numbers. In *49th AIAA Aerospace Sciences Meeting including the New Horizons Forum and Aerospace Exposition* (p. 1255). https://doi.org/10.2514/6.2011-1255

Self-Swarming for Multi-Robot Systems Deployed for Situational Awareness

Fabrice Saffre, Hanno Hildmann, Hannu Karvonen, and Timo Lind

Abstract Machine-based situational awareness is a key element to conscious and intelligent interaction with the complex world we live in, be it for the individual unit, a complex dynamical system, or even complex systems. To create this awareness, the frequent gathering of accurate and real-time intelligence data is required to ensure timely, accurate, and actionable information. Unmanned Aerial Vehicles (UAVs) and other semi-autonomous cyber-physical systems are increasing among the mechanisms and systems employed to assess the state of the world around us and collect intelligence through surveillance and reconnaissance missions. The current state of the art for humanitarian and military operations is still relying on human-controlled flight/asset operations, but with increasingly autonomous systems comes an opportunity to offload this to the devices themselves. In this chapter, we present a principled and expandable methodology for evaluating the relative performance of a collective of autonomous devices in various scenarios. The proposed approach, which is illustrated with drone swarms as an example use case, is expected to develop into a generic tool to inform the deployment of such collectives. It is expected to provide the means to infer key parameter values from problem specifications, known constraints, and objective functions.

Keywords Autonomous systems · Situational awareness · Collective intelligence · Stigmergy · Performance analysis

F. Saffre (✉) · H. Karvonen
VTT Technical Research Centre of Finland Ltd., Espoo, Finland
e-mail: fabrice.saffre@vtt.fi; hannu.karvonen@vtt.fi

H. Hildmann
TNO Netherlands Organisation for Applied Scientific Research, Den Haag, The Netherlands
e-mail: hanno.hildmann@tno.nl

T. Lind
VTT Technical Research Centre of Finland Ltd., Oulu, Finland
e-mail: hannu.karvonen@vtt.fi

T. Lipping et al. (eds.), *New Developments and Environmental Applications of Drones*,
https://doi.org/10.1007/978-3-030-77860-6_3

1 Introduction

Much has been written about the importance of accurate information, considered by some to be *"power and currency of the virtual world we inhabit"* [11]. There are classes of applications where the availability of timely, accurate, and actionable information is a key prerequisite to a successful handling of the situation. In the (civilian) context of civil security and public defence, this is the case, for example, for large events or events where complex interactions between participants can lead to problematic behaviours (crowd control [48], evacuation management [32]), or in the aftermath of a natural disaster (earthquake, flooding, large forest fires, etc. [1]).

In the military domain, the need for high-quality information sources is found in virtually all aspects of operations. *Drones*, or Unmanned Aerial Vehicles (UAVs), have been used as civilian or military surveillance tools, to patrol borders for trespassers and smugglers or to watch for enemy infiltrations [31].

1.1 Civilian and Military Use of UAVs for ISR

The umbrella term applied to all intelligence functions for military operations is *Intelligence, Surveillance, and Reconnaissance* (ISR), which originally was performed by humans, but for a number of reasons (reliability, timeliness, consistency, etc.) this is sub-optimal. The observation of an expansive geographical area over an extended period of time [36] is almost certainly going to be problematic [43]. This could be because it is uneconomical to station a dedicated human force and keep it supplied in a remote, hard-to-reach place, or because environmental conditions are harsh or hazardous, making it a difficult assignment.

Polar [9] or tropical environments, e.g., are notoriously hard to navigate for humans, and UAVs have been used for data collection in such environments with outstanding results [23]. In addition, there are dedicated wildlife areas [33] where, in addition to the difficult accessibility of the terrain, restrictions apply to the presence of humans. Yet precisely because they are remote, many of these regions are crossed by international borders or are home to vulnerable ecosystems, both of which make constant and diligent monitoring an obvious requirement. This creates the almost perfect use case for autonomous drone-based surveillance [31].

1.2 Using Autonomous UAVs as Well as Swarms Thereof

For Intelligence, Surveillance, and Reconnaissance (ISR) and, in general, for decision support, information is commonly aggregated from many different sources [47], and UAVs are ideally suited for this task. With increasing levels of autonomy becoming achievable, thanks to progress in Artificial Intelligence (AI) and related fields, missions that could conceivably be carried out by UAVs without any human intervention are rapidly growing in number. Of particular interest are self-organizing [10] groups of autonomous UAVs (drone *swarms*) that could be deployed concurrently and act as a team in the pursuit of complex, abstractly defined objectives [3]. One promising field of application for this fast-maturing technology is ISR, or *surveillance* in the civilian domain [21]. There, the use of collectives of cyber-physical systems (e.g., UAVs) operating as a single unit is increasingly considered [22]. Within ISR, there is a strong focus on mission planning and scheduling for various types of assets, such as, e.g., ground-based units, UAVs, or satellites [14]. The literature on collaborative multi-robot systems [7, 12, 24, 25] used as Mobile Sensing Platforms (MSPs) is growing fast [15, 17, 21, 22], with sub-areas developing for complex problems such as task allocation [28], multi-robot task allocation [7], group formation [42], and, more generally, self-organization [24].

Overview

Section 2 briefly discusses our stance on the use of UAVs as MSPs and provides some examples for the uses of drones at the Technical Research Centre of Finland (VTT) and the Netherlands Organisation for Applied Scientific Research (TNO). We argue for taking a biologist's view on UAV swarms (including a warning) and then elaborate on a number of aspects and challenges inherent to Multi-Robot Systems (MRSs)/Multi-Agent Systems (MASs). In Sect. 3, we define a variation of the self-swarming for Situational Awareness (SA) application (the problem) as well as propose a solution for it. Before evaluating our approach, Sect. 4 details the models used, the performance measures, and the implementation/methodology used to generate the results. With this in place, Sect. 5 provides comparative results as well as a discussion thereof.

The interested non-technical reader may immediately want to skip ahead to the conclusion (Sect. 6) where we hope to provide a concise summary of our work, situate it inside the application landscape, and, building on this, provide an outlook over how this (theoretical result) will be used in future projects.

2 Background

VTT provides a wide set of services [30, 44] and solutions [27] to UAV systems and has deep knowledge of modern UAV components, such as batteries, materials, and sensors (Fig. 1). Several next-generation new autonomous drone use cases are hosted by VTT, including projects in the areas of 5G and cyber security. At TNO, research in drone technology started as early as 1937 (see Fig. 2, courtesy of TNO Museum Waalsdorp[1]) and today, dozens of research groups across all units use drones.

[1] https://www.museumwaalsdorp.nl/en/radiocommen/telecommunication-remote-control-of-a-plane-1938-1940/.

Fig. 1 A TNO reconnaissance drone departing for automatic threat evaluation for border security and surveillance as part of the (now concluded) EU funded ALFA (Advanced Low-Flying Aircrafts Detection and Tracking) project [46]

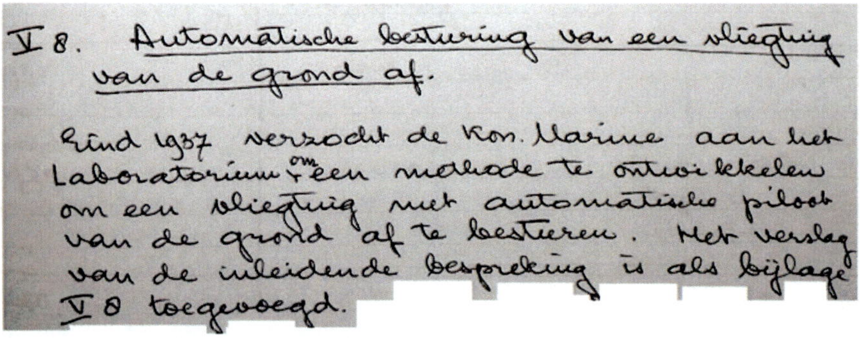

V 8. Automatische besturing van een vliegtuig van de grond af.

Eind 1937 verzocht de Kon. Marine aan het Laboratorium om een methode te ontwikkelen om een vliegtuig met automatische piloot van de grond af te besturen. Het verslag van de inleidende bespreking is als bijlage V 8 toegevoegd.

Fig. 2 TNO has been active in drone research for more than 80 years. Shown: entry (in Dutch) in the reconstructed lab records by Van Soest in 1947 and archived at TNO Museum Waalsdorp. Translation (by Eric Luiijf, TNO): "At the end of 1937, the Royal Dutch Navy requested the lab [TNO] to develop a method to control a sea plane [. . .] from the ground"

2.1 UAVs as Mobile Sensing Platforms

The usage of UAVs as MSPs [5, 15, 16, 20, 21, 26, 33] or as Wireless Senor Network (WSN)-nodes [17, 41] is growing in popularity in the literature. With regard to single UAV usage, in Fig. 1, a drone is departing for a Beyond Line of Sight (BLOS) flight [46], and Fig. 3 shows a drone performing a geological survey [40].

Due to the recent advances in the corresponding technologies [41], decreasing unit cost is making the operation of collectives of UAVs increasingly feasible [7], with Search and Rescue (SAR) operations being one of the dominant application domains for UAV swarms [24, 37, 45]. Legal restrictions [26] still make the

Fig. 3 UAV-use for geology [40]. The corresponding report can be found in [29]

Fig. 4 A field demonstration (conducted outside of the civilian airspace) for the SWACOM (Swarming and Combat Management, a project by Thales Nederland B.V., TNO-DSS, and branches of the Dutch Military) project, where multiple heterogeneous devices collectively find, and identify, objects that can pose a threat to either units in the field or civilians

practicality of operating a drone swarm in civilian airspace very difficult, but for projects such as SWACOM[2] we have access to non-civilian airspace (see Fig. 4).

2.2 *The Swarm Is More Than the Sum of Its Drones*

The argument for the use of UAVs no longer needs to be made; the literature speaks for itself, e.g., [4, 32]. With advances in the related technologies, the use of multiple UAVs as a single, physically disjunct, unit is increasingly considered in the literature, e.g., [2, 19–22, 38]. Using a term borrowed from biology, such collectives are commonly referred to as a *swarm* [20].

[2]https://www.youtube.com/watch?v=epCXIYpMSFw&t=13s.

UAV swarms offer a number of advantages over the use of a single, larger, more costly, UAV [19] ranging, e.g., from overcoming physical challenges (the curvature of the earth restricts the communication between a ground station and an aerial unit, and this is less so the case between two aerial units), operational bounds (multiple drones can replace their peers when these need to land for charging) over practical considerations (such that smaller units are harder to spot for the enemy) to financial constraints (smaller drones cost significantly less, both in Capital Expenditure (CAPEX) as well as in Operational Expenditure (OPEX)).

It should be noted that (as always) there is no proverbial *silver bullet*. UAV swarms, while offering many advantages, also have the potential for poorer results [25], if their design is not done well. Therefore, the authors would like to offer words of caution in regard to the use of swarming and other phenomena observed in biology: whenever considering such concepts from the field of biology, the practitioner should bear in mind that any such phenomena exist for a purpose (having evolved through selective processes to come into existence in the first place). Unless we (a) understand this *raison d'être* and do so (b) within the appropriate context and environment, designing concepts from biology into cyber-physical systems is merely an act of doing things for the sake of doing them. Particularly with regard to swarming, the ability for group of devices to exhibit coordinated movement through collective decision [13] has occasionally been regarded as its own reward [6], notwithstanding its usefulness in a practical application scenario.

2.3 Digital Twins

The approach used in this chapter is inspired by the way in which social insects use their environment as a shared memory. Since our locations do not actually emit pheromones, we keep track of this in a digital representation of the environment. We use a centralized memory (in the control centre) to mirror the corresponding effect a drone's path would have on the environment. In other words, we maintain a computer-based model of the real world, a so-called Digital Twin (DT).

DTs [34] are essentially virtual (thus, digital) replicas of physical systems or environments [8], with maybe the most famous example being NASA's replication of the Apollo 13 capsule on earth to assist the astronauts in space looking for a solution, given their specific circumstances and available resources. The use of DTs has become widely popular in the industry in recent years [18], especially in the context of real-time prediction, optimization, monitoring, controlling, and enhanced decision-making capabilities [39]. They are considered a technology trend and a disruptive engineering approach [34], not the least due to their potential to effortlessly integrate data (bi-directionally) between the real (physical) and the digital (virtual) version or model of a machine [18].

3 Describing the Problem and the Solution

As discussed, real-time data collection capability can be a critical factor for many applications [33], be it in the civilian [12] or in the military domain [35]. In this chapter, we propose a generic and theoretical solution to an abstract problem. As such, the presented approach constitutes the first step towards a more domain, and application, -tailored solution (see Sect. 6 for future work).

3.1 *Problem Definition and Specification*

The problem falls into the category of MRS task allocation [7, 12, 24, 25], specifically the continuous assignment of sets (sequences) of tasks (locations) to members of a swarm, over time. A collective of cyber-physical systems (a UAV swarm) is tasked with providing information about a large area; their performance is assessed through a *freshness*, summed up for all locations under surveillance.

The problem is kept generic; we distinguish the following problem-defining characteristics, expressed formally through a number of tunable parameters:

- **The environment** is reduced to three parameters or functions, namely (a) the number of unique locations in the environment, (b) how these locations are connected to one another, and (c) the physical distance between the locations.
- **The bases** where the agents (UAVs) can recharge their batteries. Both, their number and location are relevant (compare e.g., the results in Figs. 8 and 9)
- **The drones** are defined by their number, a function assigning them to starting bases and their maximum range (referred to as the *autonomy* value).
- **Permissible actions**, which in our case are simply *being present* at a location, which have the effect of resetting the signal intensity there to zero.

3.2 *The Objective*

The objective is to identify and fine-tune a local (per drone) decision-making algorithm so that the resulting collective swarm of drones exhibits collective artificial intelligence as a self-organizing swarm. The main difference with related work on this topic is the scale of the desired response, both in time and in space.

The goal is not to elicit a collective motion pattern such as flocking or schooling in a small environment (where the agents or particles operate in close proximity and directly interact with each other through, e.g., collision avoidance and/or trajectory alignment mechanisms) and over a short characteristic time scale (seconds or minutes). Instead, our objective is to develop a method that allows a swarm of drones to perform a (set of) task(s) collectively and autonomously in a very large space (square kilometres) and for an unlimited duration (weeks, months, or years).

While we use the distance measure of kilometres, the dimensions of the environment scale (up or down) with the performance of the devices as well as the size of the swarm. While the use case discussed is the surveillance of a national park using large drones, this could equally be applied to, e.g., the management of a commercial port (Rotterdam, Hamburg, etc.) using smaller drones (or autonomous surface vessels).

Within that, the performance of the swarm has to be maximized collectively, through implicit coordination between individual units. For example, two drones patrolling the same area at the same time are sub-optimal, and we want to reduce this using as little inter-drone communication as possible. In addition, the work is intentionally kept generic, but this gives rise to a number of sub-problems: navigation in an arbitrary topology, information exchange between drones, response to faults, and other unexpected events, etc.

3.3 The Approach

The core concept is that devices plan their paths individually, and based exclusively on their own perception of the environment. Taking inspiration from biology, the underlying view taken is that each location emits a signal that, as it accumulates, can diffuse into neighbouring locations. The signal intensity represents the age of the information about the corresponding location and is reset to zero when a drone visits. However, since this signal has no physical existence in the real world, the system (the set of all bases) collectively maintains a shared memory (of signal levels), to which any drone can write/from which it can read, but only *while at a base*. Therefore, the drones operate in a way that somewhat resembles the behaviour of certain social organisms such as honeybees which share information about their environment with their peers when gathering inside the hive.

Coordination between drones is achieved indirectly, through what amounts to a DT: a real-time simulation of the monitored area in which the production and diffusion of a virtual signal can be used to influence the collective response of the swarm. The underlying assumptions are that: (1) this simulation is running continuously and (2) the current state of the world's DT is accessible by drones at every base (so it can be used for path planning). This is not particularly difficult by achieve, but it requires two-way communication between the bases, so that the overall picture (local intensity of the signal at all locations) remains up-to-date throughout the system. The signal differs from what is usually called a *pheromone* in that it is not generated by the agents themselves. On the contrary, the signal is produced at a fixed rate at every location in which none is present.

3.4 The Solution

We propose a solution in which each drone determines for itself where/when to go. It does so based on model maintained by the DT and while landed at a base.

This means (a) that drones determine their next path in advance (always while landed and before they take off again) and (b) that this path is constructed based entirely on the information about the world, as maintained by the DT. Before taking off, the UAV 'flies' its intended path within the DT, which updates its model *as if the drone had already completed its journey in the real world.*

The main advantage of this approach is that it does not require any communication between units—or between units and bases—during the flight. The main drawback is that once a route has been selected and the drone has departed, it can no longer be modified through exogenous means. The control loop—executed by each drone—distinguishes a number of stages:

1. **At base and charging:** Drones will charge until their battery is full.
2. **Charged and ready:** After reaching full charge, a drone enters a *resting* state out of which it will come with a fixed probability per predefined time period.
3. **Planning a route:** A UAV has a maximum range (the number of transitions from one location to another in a discretized world map which a drone can complete with a single battery charge). Planning happens in two stages:

   (a) *Outbound:* As long as the battery charge (remaining hops) is larger than the shortest path (number of hops) from the current location to the nearest base, the drone may select any neighbour of its current position as the next location to visit. This decision uses the signal levels of the neighbouring locations, either deterministically (picking always the one with the highest intensity) or stochastically with a variable non-linear component (the stronger the signal from a location, the higher the chance of it being chosen as next hop). The current implementation uses the deterministic variant, with investigations into the performance of a stochastic choice left for future work (see Sect. 6).

   (b) *Inbound:* Once the distance to the nearest base reaches the same value as the remaining autonomy (battery charge), the UAV becomes restricted in the choice of its next destination. In short: it uses the same method to determine the next hop as before, but this time only those locations that reduce the number of hops to a base are considered. NB: This means that a drone can return to the base from which it started or head for a different one.

4. **Departure:** Before the UAV departs, it updates the DT which treats the plan as *fait accompli* and resets the signal intensity for all locations in the flight path to zero. The drone then performs its flight autonomously and only reconnects to the system after it has landed at the next base.

Resetting the intensity of the signal in the DT immediately along the entire route prior to the drone's physical departure has two advantages: (1) it makes it easier to implement, since all updates are made in one go and without the need for any

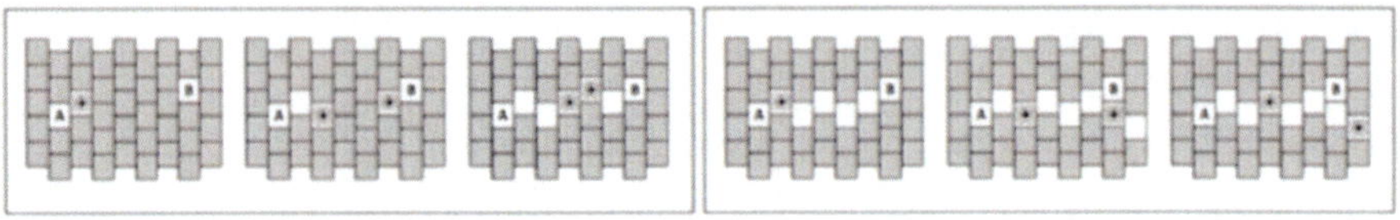

Fig. 5 Illustrating the benefit of *immediately* resetting the signals along the entire path (right image), as opposed to doing so per location, when visiting that location (left image). The first drone starts in base A and is planning to move to base B. In the left sequence, signals are only removed from a location when the drone moves there. In this situation, a second drone (in base B) may plan the exact same route—only in reverse (from base B to base A), which is effectively redundant

interaction with airborne units and (2) it means that distant waypoints along the route become less attractive to other drones planning their own patrol immediately (i.e., before they are even reached by the departing unit). Point (2) tends to limit duplication of effort by reducing overlap between waypoint sequences. The idea behind the approach is briefly illustrated in Fig. 5 on the next page.

4 Materials and Methods

4.1 The World

The map is discretized using a hexagonal mesh. This choice ensures that the distance between any two neighbours (candidate waypoints) has the same value for all connections (in our simulations this value is set to 1). A distance of 1 can be understood as 100 m to allow some comparison to the real world.

Maps are generated by creating the hexagonal mesh equivalent of a 5 km square (25 km^2), fully connected, and then iteratively removing nodes randomly, starting from the outer edge (i.e., only nodes with less than 6 neighbours can be removed) to generate a realistically complex geographical area to *patrol*. A sample of 1000 such regions was created and used consistently throughout all simulations. The final sample was comprised of 995 regions (5 had to be discarded due to their being disconnected) varying in size from just above 2 to just under 21 km^2. Figure 6 shows two representative examples of a typical layout.

The assets, i.e., drones and drone bases, were set as follows: unless stated otherwise, one base is created per km^2 (rounded down), so the number varied between 2 and 20 bases for the aforementioned sample. With regard to the actual placement of the bases, two variations were implemented: (1) purely random and (2) *near optimal*, meaning that starting from random locations, bases were moved apart following a repulsive field until an approximately even distribution (within the confines of the region of interest) was reached. Unless stated otherwise, the number of drones is equal to the number of bases and their initial location (at which base they start) is chosen at random (so it is possible for two drones to start at the same base).

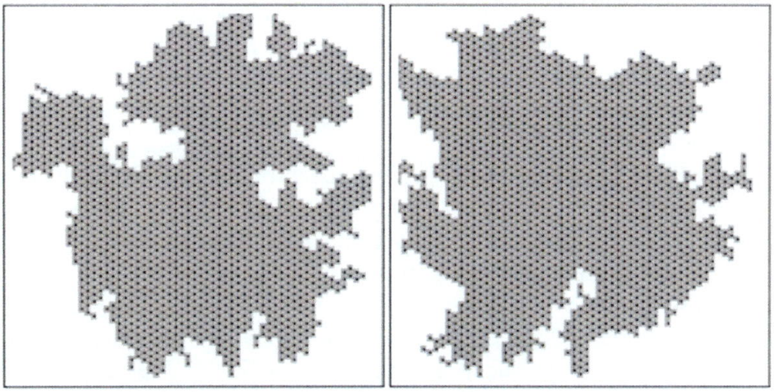

Fig. 6 Two examples of a hexagonal mesh (meaning that any node has exactly 6 possible neighbouring locations to which it can be connected) superimposed over an arbitrarily shaped geographical area. In both cases, the square in which the swarm's territory is inscribed is 5×5 km (for a mesh in which any two connected nodes are 100 m apart)

With regard to power consumption and battery charging, each hop incurs a cost of 1, while each time step spent at a base recharges the battery by 1. The autonomy of each drone (at full charge) was set to 60. If considering the 100 m distance between waypoints, this corresponds to a total distance travelled per trip of 6 km. If the drones' autonomy is a realistic 30 min, this in turn implies a time unit of 30 s and a flight speed of 12 km/h. NB: These values are approximations and are only intended to give the reader a rough idea of the scale at which the proposed system could operate.

4.2 *Departure and Path Planning*

The probability to leave the *resting* state after reaching full charge was set so that, statistically, each drone would spend the same time flying, charging, and on stand-by, which in turn implies that about one-third of the fleet can be expected to be airborne at any one time. Every drone computes its flight plan based on information available from the DT at the time of departure, as per the procedure detailed in Sect. 3.4.

4.3 *Performance Measure*

For the performance measure, we need to reiterate the intended application domain, namely the *continuous* allocation (observation) of locations to schedules such that we minimize the age of the information as much as possible. This value is already recorded for all locations in the DT. This allows us to determine, for any number of time steps, how many locations have not been visited during that period. Conversely, for any fraction of locations, we can determine how many steps (on average) it takes such that all locations have been visited at least once.

For example, if there are 6 drones flying concurrently and 120 nodes (roughly covering one square kilometre at 100 m interval), at least 95% (114) will not have been visited less than one time step ago (possibly more if two or more drones happen to be co-located). If this threshold is increased to one (visited less than two time steps ago), this fraction could drop at most to 90% (108), since, in the absence of any overlap, both the nodes closest to the drones now and those closest to them one time step ago will be excluded (and so on and so forth). The rate at which this fraction of "surviving" nodes decreases as the value of the threshold increases is a good indication of performance for the chosen objective of effectively patrolling the area. For example, it is reasonable to say that a distributed algorithm for which, at any point in time (after the system has reached steady state), 50% of the nodes will have statistically been visited in the last 100 time steps is better than one for which reaching the same fraction (0.5) requires extending the threshold to 150 time steps. Furthermore, this measure can easily be extended to include a wider area around the drones, using a revised formulation such as *"what fraction of nodes have not been within 1, 2, ..., n hops of any drone in the last t time-steps?"* (see Fig. 7).

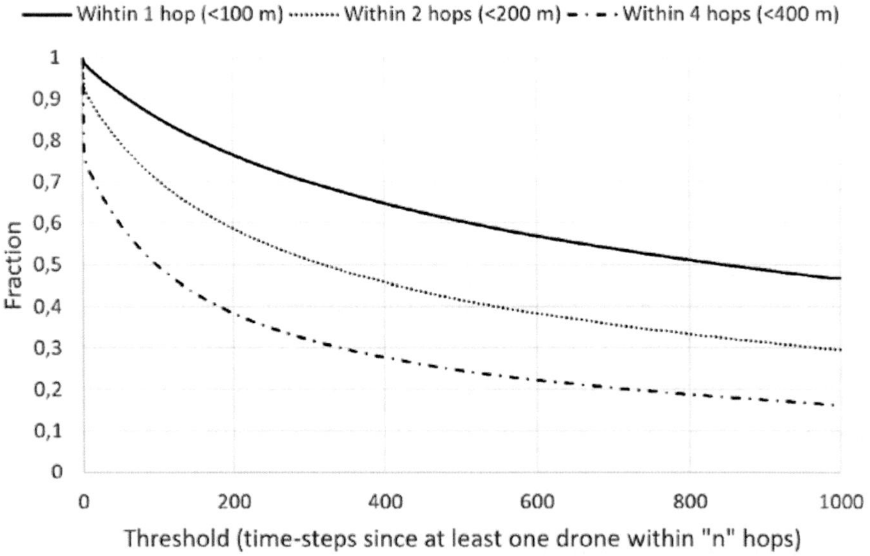

Fig. 7 The performance metric. The *survival* curves indicate the fraction of nodes in the hexagonal mesh (vertical axis) that have not been within 100, 200, and 400 m of at least one drone in the last t time steps (horizontal axis) i.e., 59% of the area was always beyond one hop of the nearest drone in the last 200 time steps, 38% if extending the time window to 600 steps (middle curve)

4.4 Data Collection

Two hundred independent simulation runs (including the initialization phase in which the base and drone locations are determined) were conducted for every region in the sample. So for every combination of parameter values, results are compiled from nearly 200k realizations. Each simulation lasted 2000 time steps, which was found to be sufficient for the system to have reached steady state.

5 Results and Discussion

The results are the summary of more than 2 million independent simulations, with the tested scenarios being imaginary maps (i.e., not real locations); the results and the discussion are to be seen in this light.

5.1 Benchmark Performance Analysis

As a benchmark, we considered randomly placed bases. As previously stated, 995 simulated environments of varying areas were used, and the presented results are the average values from 200 individual runs for each combination of parameter values. Figure 8 plots for each of the tested environments the average number of steps it took (y-axis) to achieve one of the three measures of, basically, complete coverage, as a function of the total area of the region of interest (x-axis). Recorded is the time taken to visit 99% of all locations, defined using the increasingly inclusive criteria of having been within 100, 200, or 400 m of at least one drone.

It is important to keep in mind that, unless stated otherwise, the ratio between the number of drones and the area size remains the same for all simulations (so for significantly larger environments, there is also a significantly larger sized swarm). The fact that the approach actually appears to perform better for larger environments is therefore not surprising: the law of big numbers dictates that it will be increasingly unlikely to have all drones (and bases) placed such that some part of the environment remains *unvisited* for a long period.

It is also important to consider the following: the decision to measure when a location was visited, but also when a location was almost visited is motivated by the nature of the approach. We use signals to guide the drones' paths but restrict their movement to a hexagonal mesh. In this constellation, we can easily come up with examples where visiting all locations comes at a high cost (i.e., has to go through locations that have recently been visited). We argue that this measure is appropriate for the applications we have in mind, which consider the drones carrying advanced sensing equipment that can operate over distances and thus assess the condition of neighbouring locations.

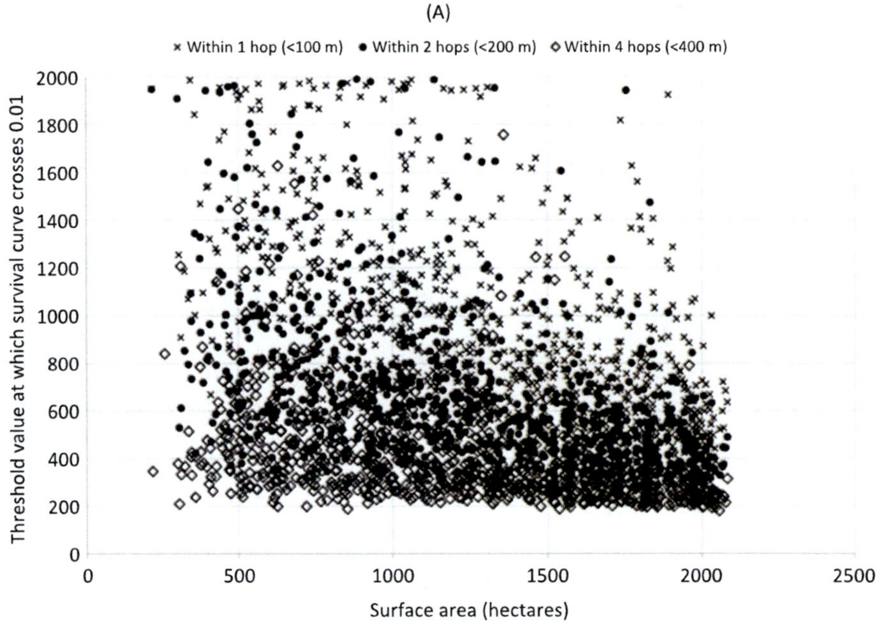

Fig. 8 Results for starting results for starting with a single drone per base and a fixed "number of bases to mapsize" ratio. 995 simulated environments were evaluated (plotted along the x-axis on the basis of their size, not topology). Recorded are the average performance (over 200 simulations, see Figure 10 for the variance) with regard to the time taken (y-axis) to collect data from 99% of the locations, where the data collection capability was assumed to be such that cells within 1, 2, or 4 hops could be observed. Figure 9 shows results for twice the number of bases or twice the number of drones

5.2 *Influence of the Number of Bases and Drones*

In the benchmark, the ratio between the number of bases and the size of the area was kept constant, with the number of drones equal to the number of bases. This was to quantify the influence that the size of the region of interest has on performance, for a fixed average population density. The obvious next step, particularly with respect to identifying parameter values capable of delivering a target performance in a future real-world deployment, is to relax this fixed population density rule.

We started by examining two additional scenarios: one in which the number of drones was doubled but the number of bases was kept identical (relative to benchmark values), and vice versa (twice as many bases, same number of drones as in the benchmark). The results are summarized in Figs. 9 and 10.

Unsurprisingly, increasing the number of drones resulted in a marked performance increase compared to the benchmark. From a qualitative viewpoint, this is trivial (having more units can only reduce the time interval required to reach any given level of coverage), but it is informative to compare the gain achieved (as

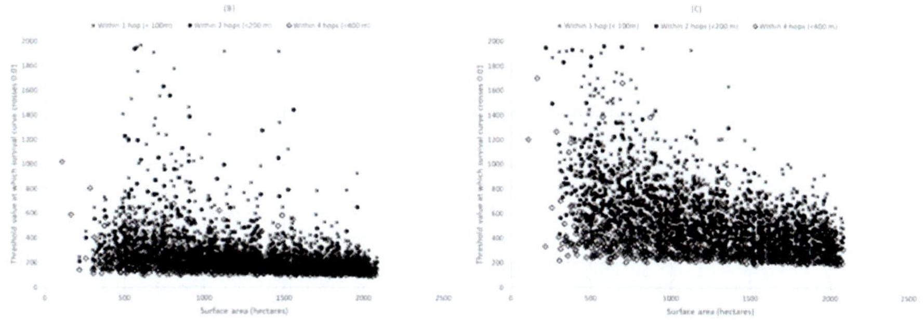

Fig. 9 Identical plot as in Figure 8, but for double the number of drones (left) or double the number of bases (right). Bases are distributed randomly

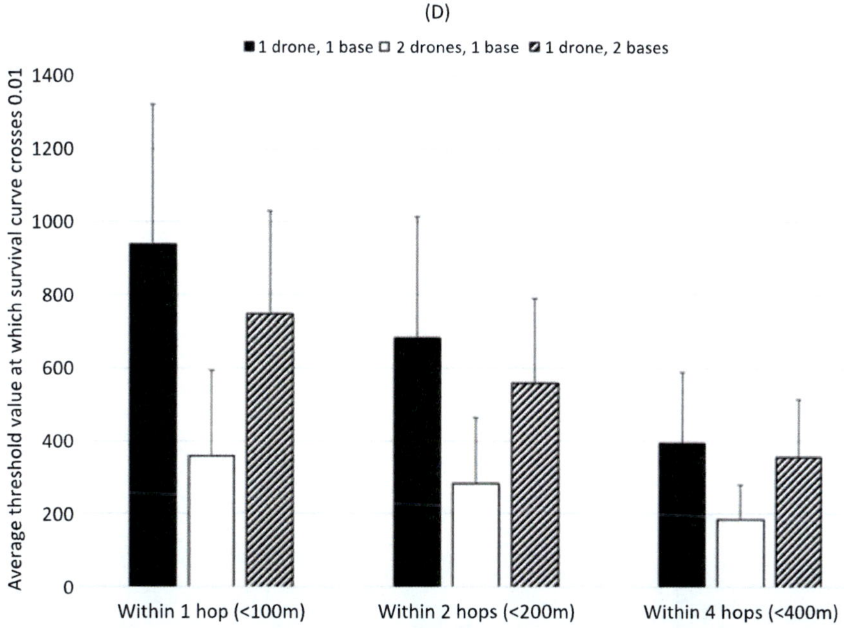

Fig. 10 Performance comparison between the benchmark scenario (full bars), the modified version involving twice the number of drones (empty bars), and the one involving twice the number of bases (dashed bars). The chosen metric is the average value of the threshold time interval for the three criteria. Error bars indicate the standard deviation

a function of the severity of the criterion) quantitatively. For the least inclusive one (<100 m), the performance increased by 62% on average, whilst for the most inclusive (<400 m), this figure dropped to 53% (see Fig. 10).

The effect of increasing the number of bases is more interesting and less intuitive: doubling this parameter value appears to result in a significant improvement, reducing the threshold value by an average 20% for the toughest criterion. This is easily explained by more patrol routes becoming available as the number of bases increases, but since the cost of these charging stations is likely to be substantially lower than that of a drone, it also strongly suggests that this approach could be a more efficient strategy to achieve a given target performance.

5.3 The Impact of Base Locations

A final set of numerical experiments involved a preliminary exploration of the influence of base placement. The fact that performance generally increases with the size of the region of interest when the number of bases is linearly proportional to the surface area (benchmark scenario) seems to indicate that poor placement could have a strong negative impact. Accordingly, we tested an alternative scheme in which bases are placed so as to approximate an even distribution (see Sect. 4.1 for details).

Results are shown in Fig. 10 and confirm that this modification had a beneficial effect, although it was not as pronounced as anticipated. We believe that this is due to the relatively long range of the drones relative to the typical environment size. This tends to limit the impact of the charging points location, since it makes it less likely that any part of the region of interest is beyond reach, even if it is far from the nearest base. We will seek to confirm this interpretation in future work (Fig. 11).

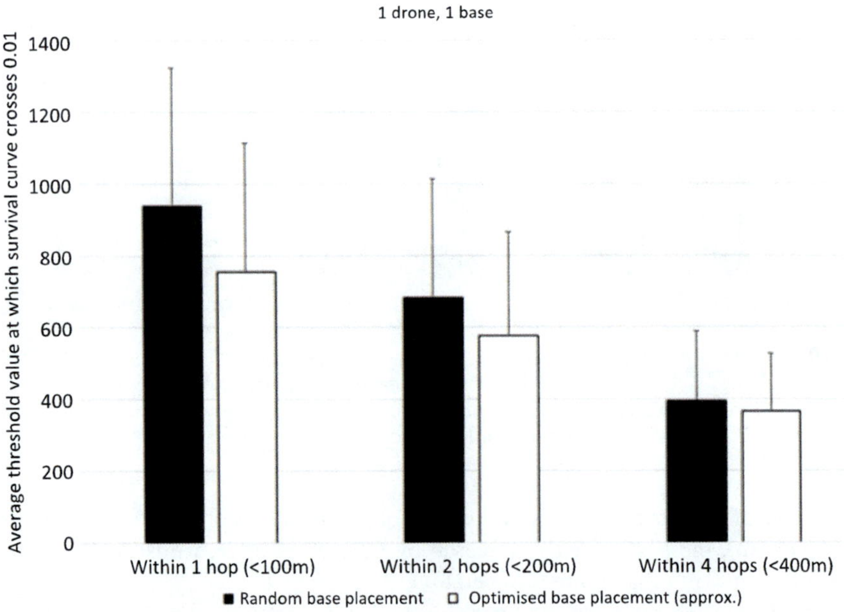

Fig. 11 Comparison between random (full bars) and near-optimal (empty bars) base placement

6 Conclusion

The considerations presented in this chapter lay the ground work for general investigations into path planning for data collection of a collective of cyber-physical systems. Both VTT and TNO engage in this theoretical research for reasons related to a large number of projects, involving numerous types and classes of unmanned and potentially autonomously operating vehicles. Applications range from assessing conditions in the environment in real time (monitoring air quality, looking for gas leaks, surveillance for large industrial terrains) over military intelligence gathering (finding land mines, tracking enemy movement, and ultimately engaging in armed conflict) to, in the future, deployment of intelligent autonomous systems over extremely large distances (in space or on other celestial bodies). All these applications are currently being worked on in some way or the other, and many of these use cases will become reality in the years to come.

6.1 Scope and Applicability of the Presented Idea

The title of this chapter intentionally speaks of *multi-robot* systems, as opposed to restricting the applicability of the presented work to swarms of UAVs. Within the scope of the presented work, one might argue the presented approach and the discussed results are more applicable to, e.g., ground-based drones such as rovers because the provided modelling glosses over a number of issues that UAVs would face for the practical implementation. For example, the range of a UAVs depends on the wind conditions at the time, and due to this, connections between two locations do not incur symmetric energy costs (unlike in our model).

The specific requirements of any use case will determine how, and to which extent, an application can be tuned and tailored. Any expert and practitioner in the field knows that after a system is built, optimizing it for the specific use case will often hold the potential for significant performance improvements and will, quite commonly, be what can make the difference between outperforming the competition or being outperformed. With that in mind, we proposed a straightforward, simple mechanism that enables a collective of devices to operate on a shared problem. The approach side-steps all difficult inter-agent negotiation processes (undoubtedly at some cost to the performance) and, due to its simplicity, is inherently robust against unit failure. The need for secure communications is removed, as the devices do not need to communicate for the planning and execution of the plan (drones may still breaking radio silence when they have spotted something noteworthy).

6.2 Future Work

We randomly created nearly a thousand maps of different sizes and extensively tested various settings for all of them. Even though none of these represents a real-world location or use case, clear trends can be identified. For instance, for constant surface area/swarm size ratio, performance improves as the environment becomes larger. Such results can and will inform current and future projects where case-driven approaches have to be designed and implemented. Based on the current projects, and those planned/expected in the near future, we consider the following as *future work*.

6.2.1 Realistic Cost Functions

Realistic cost functions for calculating and planning the routes. This means including relevant environmental conditions such as wind speed and direction, but also using different drone models. A rover on Mars, for example, can simply pause operations and resume them when conditions have improved while a UAV over a battlefield cannot. These investigations will be project-driven and probably happen in parallel to investigations of using heterogeneous collectives (either in the equipment the drones carry (not everyone has a temperature sensor) or in their type or class, e.g., using rovers and drones together).

6.2.2 Increasing Agent Autonomy

VTT and TNO are working toward increasing agent autonomy in the sense of endowing the drones with more advanced intelligence and decision-making capabilities. Our drones can already carry significant payloads with regard to sensing, as well as data processing, hardware. In the various scenarios presented in this chapter, a drone simply follows its planned route, but it seems obvious that in any realistic deployment, future surveillance drones would need to be capable of deviating from their course in response to detecting certain events. Future work will investigate the interplay between the type of orchestration methods described here and the situational awareness and other autonomous features being developed in a variety of other projects at TNO and VTT.

6.2.3 Adding Priorities

Adding priorities to be able to have a differentiated view on the map. This has a connection to the previous point as drones could change the priority of a location based on data that they have collected during their journey. We are currently considering local alterations to the signal production or diffusion rate as a convenient route to implement such variable priorities.

6.2.4 Additional Investigation of Topology and Swarm Size

We want to continue our systematic exploration of how swarm size impacts performance, and what practical implication this can have (consider, for example, the possibility of altering the number of drones post-deployment, based on information collected at runtime). Furthermore, we are in the process of evaluating a measure of complexity for the generated maps (fractal dimension) that would allow for a more fine-grained investigation of the influence of the region's characteristics than simply considering the total surface area. If successful, applying the same metric to any real map could allow us make statistical predictions about the performance of any given deployment of the proposed solution (swarm size, drone range, number of bases, DT parameter values, etc.).

6.2.5 Stochastic Versus Deterministic Planning

In the variation currently used, the paths are planned deterministically based on the signals in neighbouring locations. We are currently investigating a stochastic alternative, which could result in additional flexibility.

Acknowledgments This research was partly supported by VTT grant 127686 (`GG_SelfSwarm`). The Academy of Finland is acknowledged for financial support of project "`Finnish UAV Ecosystem`" 337878 (project decision number 337878). This research was also partly funded by the Scientific Advisory Board for Defence (MATINE) in the `CounterSwarm` project. The picture of the diary, used for Fig. 2, was kindly provided by Eric Luiijf and the TNO Museum. The authors wish to thank Janne Järvinen, Matthieu Molinier, and Tuomo Tuikka for useful discussions and suggestions.

References

1. Al-Kaff, A., Madridano, Á., Campos, S., García, F., Martín, D., & de la Escalera, A. (2020). Emergency support unmanned aerial vehicle for forest fire surveillance. *Electronics, 9*(2), 260.
2. Al-Naji, A. A., Perera, A., Mohammed, S. & Chahl, J. (2019). Life signs detector using a drone in disaster zones. *Remote Sensing, 11*, 2441.
3. Almeida, M., Hildmann, H., & Solmaz, G. (2017). Distributed UAV-swarm-based real-time geomatic data collection under dynamically changing resolution requirements. In *UAV-g 2017—ISPRS Archives of the Photogrammetry, Remote Sensing and Spatial Information Sciences*, Bonn, Germany.
4. Alzahrani, B., Oubbati, O. S., Barnawi, A., Atiquzzaman, M., & Alghazzawi, D. (2020). UAV assistance paradigm: State-of-the-art in applications and challenges. *Journal of Network and Computer Applications, 166*, 102706.

5. Ancin-Murguzur, F. J., Munoz, L., Monz, C., & Hausner, V. H. (2020). Drones as a tool to monitor human impacts and vegetation changes in parks and protected areas. *Remote Sensing in Ecology and Conservation, 6*(1), 105–113.
6. Awasthi, S., Balusamy, B., & Porkodi, V. (2019). *Artificial intelligence supervised swarm UAVs for reconnaissance* (pp. 375–388).
7. Bahgat, A., Shehata, O., & El Sayed, M. (2020). A multi-level architecture for solving the multi-robot task allocation problem using a market-based approach. *International Journal of Mechanical Engineering and Robotics Research*, 293–298.
8. Barricelli, B. R., Casiraghi, E., Fogli, D. (2019). A survey on digital twin: Definitions, characteristics, applications, and design implications. *IEEE Access, 7*, 167653–167671.
9. Berglund, R., & Seitsonen, L. (2019). DronIce: Using drones to support icebreaker operations in the Baltic Sea. In *Living Planet symposium 2019*. Conference date: 13-05-2019 Through 17-05-2019.
10. Boubeta-Puig, J., Moguel, E., Sánchez-Figueroa, F., Hernández, J., Carlos Preciado, J. (2018). An autonomous UAV architecture for remote sensing and intelligent decision-making. *IEEE Internet Computing, 22*(3), 6–15.
11. Branwyn, G., Idol, B. (1993). Cyberpunk, UOM. Emi Special Issue ASIN: B00000IN06, BSIN: B00000IN06.
12. Conesa-Muñoz, J., Valente, J., Del Cerro, J., Barrientos, A., & Ribeiro, A. (2016). A multi-robot sense-act approach to lead to a proper acting in environmental incidents. *Sensors, 16*(8), 1269.
13. Dai, F., Chen, M., Wei, X., & Wang, H. (2019). Swarm intelligence-inspired autonomous flocking control in UAV networks. *IEEE Access, 7*, 61786–61796.
14. Duan, L., Luo, B., Li, Q., & Yu, G. (2016), Research on intelligence, surveillance and reconnaissance mission planning model and method for naval fleet. In *2016 Chinese Control and Decision Conference (CCDC)* (pp. 2419–2424).
15. Eledlebi, K., Ruta, D., Hildmann, H., Saffre, F., Alhammadi, Y., & Isakovic, A. F. (2020). Coverage and energy analysis of mobile sensor nodes in obstructed noisy indoor environment: A Voronoi-approach. *IEEE Transactions on Mobile Computing*.
16. Eledlebi, K., Ruta, D., Saffre, F., AlHammadi, Y., & Isakovic, A. F. (2018). Voronoi-based indoor deployment of mobile sensors network with obstacles. In *2018 IEEE 3rd International Workshops on Foundations and Applications of Self* Systems (FAS*W)* (pp. 20–21).
17. Eledlebi, K., Hildmann, H., Ruta, D., & Isakovic, A. F. (2020). A hybrid Voronoi tessellation/genetic algorithm approach for the deployment of drone-based nodes of a self-organizing wireless sensor network (WSN) in unknown and GPS denied environments. *Drones, 4*(3), 33.
18. Fuller, A., Fan, Z., Day, C., & Barlow, C. (2020). Digital twin: Enabling technologies, challenges and open research. *IEEE Access, 8*, 108952–108971.
19. Hentati, A. I., & Fourati, L. C. (2020). Comprehensive survey of UAVs communication networks. *Computer Standards & Interfaces, 72*, 103451.
20. Hildmann, H., Eledlebi, K., Saffre, F., & Isakovic, A. F. (2020). Internet of Drones. In *Chapter the swarm is more than the sum of its drones—A swarming behaviour analysis for the deployment of drone-based wireless access networks in GPS-denied environments and under model communication noise*. Studies in Systems, Decision and Control. Springer.
21. Hildmann, H., & Kovacs, E. (2019). Review: Using unmanned aerial vehicles (UAVs) as mobile sensing platforms (MSPs) for disaster response, civil security and public safety. *Drones, 3*(3), 59.
22. Hanno Hildmann, Ernö Kovacs, Fabrice Saffre, and A. F. Isakovic. Nature-inspired drone swarming for real-time aerial data-collection under dynamic operational constraints. *Drones*, 3(3), 2019.
23. Hodgson, J. C., Baylis, S. M., Mott, R., Herrod, A., & Clarke, R. H. (2016). Precision wildlife monitoring using unmanned aerial vehicles. *Scientific Reports, 6*(1), 22574.
24. Hussein, A., Adel, M., Bakr, M., Shehata, O. M., & Khamis, A. (2014). Multi-robot task allocation for search and rescue missions. *Journal of Physics: Conference Series, 570*(5), 052006.

25. Hussein, A., & Khamis, A. (2013). Market-based approach to multi-robot task allocation.
26. Jiménez López, J., & Mulero-Pázmány, M. (2019). Drones for conservation in protected areas: Present and future. *Drones, 3*(1), 10.
27. Karvonen, H., Heikkilä, E., & Wahlström, M. (2020). Safety challenges of AI in autonomous systems design—solutions from human factors perspective emphasizing AI awareness. In D. Harris & W.-C. Li (Eds.), *Engineering psychology and cognitive ergonomics. Cognition and design* (pp. 147–160). Cham: Springer.
28. Khamis, A., Hussein, A., & Elmogy, A. (2015). *Multi-robot task allocation: A review of the state-of-the-art* (pp. 31–51). Cham: Springer.
29. KNGMG en NWO. (2019). *Geo Brief* (Vol. 44).
30. Laarni, J., Koskinen, H., & Väätänen, A. (2017). Concept of operations development for autonomous and semi-autonomous swarm of robotic vehicles. In *Proceedings of the Companion of the 2017 ACM/IEEE International Conference on Human-Robot Interaction, HRI '17* (pp. 179–180). New York, NY: Association for Computing Machinery.
31. Long, T., Ozger, M., Cetinkaya, O., & Akan, O. B. (2018). Energy neutral internet of drones. *IEEE Communications Magazine, 56*(1), 22–28.
32. Makinoshima, F., Imamura, F., & Oishi, Y. (2020). Tsunami evacuation processes based on human behaviour in past earthquakes and tsunamis: A literature review. *Progress in Disaster Science, 7*, 100113.
33. Mangewa, L. J., Ndakidemi, P. A., & Munishi, L. K. (2019). Integrating UAV technology in an ecological monitoring system for community wildlife management areas in Tanzania. *Sustainability, 11*(21), 2019.
34. Moyne, J., Qamsane, Y., Balta, E. C., Kovalenko, I., Faris, J., Barton, K., & Tilbury, D. M. (2020). A requirements driven digital twin framework: Specification and opportunities. *IEEE Access, 8*, 107781–107801.
35. Novák, A., Bufita, L. (2019). Business intelligence and ISR data processing. In *2019 International Conference on Military Technologies (ICMT)* (pp. 1–8).
36. Papakonstantinou, N., Bashir, A. Z., O'Halloran, B., & Bossuyt, D. L. V. (2019). Early assessment of drone fleet defence in depth capabilities for mission success. In *2019 Annual Reliability and Maintainability Symposium (RAMS)* (pp. 1–7).
37. Półka, M., Ptak, S., & Kuziora, L. (2017). The use of UAV's for search and rescue operations. *Procedia Engineering, 192*, 748–752. 12th International scientific conference of young scientists on sustainable, modern and safe transport.
38. Ramirez-Atencia, C., R-Moreno, M. D., & Camacho, D. (2017). Handling swarm of UAVs based on evolutionary multi-objective optimization. *Progress in Artificial Intelligence, 6*(3), 263–274.
39. Rasheed, A., San, O., & Kvamsdal, T. (2020). Digital twin: Values, challenges and enablers from a modeling perspective. *IEEE Access, 8*, 21980–22012.
40. Reindersma, R. N. (2019). Capturing sedimentary outcrops in 3D, putting the geologists' needs into practice. Master's thesis, Utrecht University, The Netherlands.
41. Oubbati, O., Atiquzzaman, M., Ahanger, T., & Ibrahim, A. (2020). Softwarization of UAV networks: A survey of applications and future trends. *IEEE Access, 8*, 98073–98125.
42. San Frutos, R. S., Al Kaff, A., Hussein, A., Madridano, Á., Martín, D., & de la Escalera, A. (2019). ROS-based architecture for multiple unmanned vehicles (UXVs) formation. In *Computer Aided Systems Theory—EUROCAST 2019* (pp. 11–19). Cham: Springer.
43. Tuominen, S., Balazs, A., Saari, H., Pölönen, I., Sarkeala, J., & Viitala, R. (2015). Unmanned aerial system imagery and photogrammetric canopy height data in area-based estimation of forest variables. *Silva Fennica, 49*(5).
44. Väätänen, A., Laarni, J., & Höyhtyä, M. (2019). Development of a concept of operations for autonomous systems. In Jessie Chen, editor, *Advances in Human Factors in Robots and Unmanned Systems* (pp. 208–216). Cham: Springer.

45. van den Broek, A. C., Dekker, R. J.: Geospatial intelligence about urban areas using SAR. In M. Ehlers, & U. Michel (Eds.), *Remote sensing for environmental monitoring, GIS applications, and geology VII* (Vol. 6749, pp. 199–210). International Society for Optics and Photonics, SPIE.
46. van den Broek, B., van der Velde, J., van den Baar, M., Nijsten, L., & van Heijster, R. (2019). Automatic threat evaluation for border security and surveillance. In *Counterterrorism, crime fighting, forensics, and surveillance technologies III* (Vol. 11166, pp. 113–122). Int. Society for Optics and Photonics, SPIE.
47. Voogd, J., de Heer, P., Veltman, K., Hanckmann, P., & van Lith, J. (2019). Using relational concept networks for explainable decision support. In *Mach. learning and knowledge extraction* (pp. 78–93). Cham: Springer.
48. Zhang, Z., Herrera, N., Tuncer, E., Parr, S., Shapouri, M., & Wolshon, B. (2020). Effects of shadow evacuation on megaregion evacuations. *Transportation Research Part D: Transport and Environment, 83*, 102295.

Toward Invisible Drones: An Ultra-HDR Optical Cloaking System

Karri Palovuori

Abstract In order to reduce the visual detectability of drones, an active cloaking system was developed to match their color against the background sky. The system consists of an embedded control system connected to a smart LED tapestry and two color sensors, all capable of operating over an extreme dynamic range of 1 : 1 000 000. The cloaking system was applied to a commercial drone, and the results under widely varying outdoors conditions are reported. The cloaking system successfully matches all background sky conditions, save the solar disk or halo.

Keywords Drone · Visibility · Invisibility · Camouflage · Twilight · Adaptive

1 Introduction

Depending on the situation, the visibility of airborne drones can be a vice or a virtue. On one hand, to signal their presence to people underneath and to other air traffic, bright lights are added to drones for their effortless detection during dark. This is also mandated by regulations under many jurisdictions [1, 2]. On the other hand, sometimes minimal disturbance to the environment is preferred. Such cases include non-distractive observation of wildlife or livestock and covert surveillance.

The elimination of the signal lights reduces the visibility of a drone, but the drone hull is highly contrasted against the sky under any environmental circumstances, except during the nighttime. Even white drones, which diffusely reflect the average ground color, appear visually as black (Fig. 1a) in flight. During the night, any non-illuminated drone is obviously invisible by default.

To cloak a drone against the sky requires it to have the same apparent color and luminance toward the observer as the section of the sky seen surrounding it. While this fundamental principle is well known from both fiction [3] and history [4],

K. Palovuori (✉)
Tampere University, Electronics, Tampere, Finland
e-mail: karri.palovuori@tuni.fi

T. Lipping et al. (eds.), *New Developments and Environmental Applications of Drones*,
https://doi.org/10.1007/978-3-030-77860-6_4

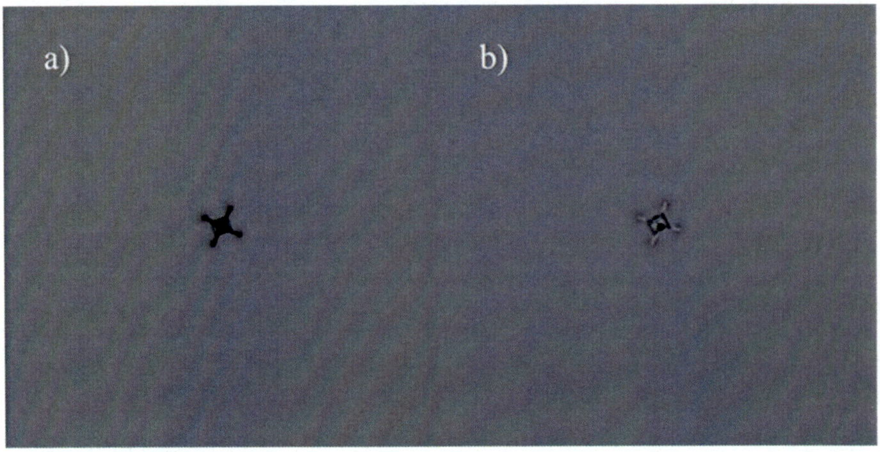

Fig. 1 (**a**) A *white* drone against overcast sky, (**b**) the same drone cloaked (the undercarriage and the camera remain clearly visible)

its application to drones presents serious technological challenges: the maximum brightness of the sky is very high, and its luminous intensity has a vast dynamic range. These are both difficult to match by an illumination system and to accurately measure by a sensor.

The clear sky at midday is blue and has a typical luminance of about 8000 cd/m^2 [5]. The minimum illumination condition under which any targets against the sky can be detected by the naked eye is defined as nautical twilight [6], during which the sky has a luminance of about 0.005 cd/m^2 [7]. These two limits suggest a requirement for the dynamic range of roughly 1 : 1 600 000 for both the measuring and the illumination systems. In contrast, the dynamic range of a typical 8-bit computer display is 1 : 255 and that of an entry-level 10-bit HDR display 1 : 1023.

If the location or direction of the observer is not known, the drone has no information of the exact direction of the sky to match its color to. In such a case, a default decision has to be made. Two obvious choices are the color of the zenith and the average color of the whole sky. Since the dynamic range of the colors the sky can simultaneously have is from near black to vivid dusk or dawn, an average is problematic and likely to yield a very incorrect result. We used the color of the zenith in all our measurements as the target color and observed the drone from underneath.

2 Methods

The cloakings system consists of the light-emitting cloak itself, two measurement systems to measure the target color and the current cloak color, and a control system to match those. The cloaking system was developed in steps, by first building a handheld test patch to verify the feasibility before integrating it to a midsized commercial drone.

2.1 The Cloak

The cloak itself, i.e., the light-emitting surface was constructed of a commercial unbranded adhesive LED strip consisting of 144 APA102C RGB LEDs per meter on a flexible PCB of 12.5 mm width [8]. This particular type was originally selected due to its high brightness, anticipating challenges in matching the brightness of the sky. The LEDs on the strip have a Lambertian radiation pattern, appearing of constant luminance from any forward viewing angle.

The APA102C RGB LEDs each have a red, a green, and a blue LED chip and a smart controller integrated into a small surface mount package. The data is transferred synchronously with a data signal and a clock signal from unit to unit, which can be daisy chained indefinitely. Each unit regenerates both the signals to the next unit with a delay of half a clock period. The data stream contains a simple synchronization pattern and after that 32 bits of control data for each unit. Each unit strips the 32 first control bits after the synchronization for its own use and then transmits the rest unmodified.

The APA102C has two simultaneous brightness controls. First, each RGB color is defined with a 24-bit word, i.e., 8 bits per component. Second, the total dimming of the LED is defined with a 5-bit word [9]. As a result, the brightness of an individual RGB LED has a dynamic range of 13 bits. With a larger number of LEDs, dithering can expand the average dynamic range greatly.

At maximum brightness each RGB LED consumes 60 mA of current resulting in total current consumption of almost 9 A per meter of strip. This is enough to generate significant ohmic losses over the power supply traces on the strip and a subsequent drop in operating voltage of the LEDs, apparent already with relatively short runs of the strip. The operating voltage changes directly affect not only the brightness of the LEDs but also their color as the different color chips have different threshold voltages. Therefore, simple feedforward control is not enough to ensure a correct outcome from this type of light source, and feedback is required. This compensates for all minor error sources (e.g., temperature, batch variation, defective chips), too.

The maximum density of the LEDs in the cloak is defined by the strip geometry. Along the strip the LEDs are placed with an interval of 6.9 mm (144 per meter), and the strip is 12.5 mm wide. The perfect visual acuity of the human eye is enough to discriminate details of ~1 arc minute [4], suggesting the individual LEDs or LED strips cannot be resolved when viewed from further than ~ 40 m. In practice, it is unlikely that the viewer would observe a cloaked target with foveal vision and perfect focus, before it is visually acquired. Thus the cloak is possibly effective from shorter distances, too.

2.2 *The Measurement System*

In order for the cloak to match the color of the sky, the sky color needs to be measured. The dynamic range requirement of this measurement system is very large compared to common optical sensors. At the low end, cameras report colors with 8 bits of resolution per component and even dedicated color sensors only with 16 bits. The sensor of choice was the VEML6040 color sensor, which has a resolution of only 16 bits but in addition has a configurable integration period directly affecting sensitivity, expanding the total dynamic range to 21 bits [10]. The sensor has an I^2C bus for communication.

The sensor chip itself has a Lambertian sensitivity pattern, and to measure the color of the sky from a particular direction, the field of view of the sensor needs to be restricted. A baffle for this purpose was designed, 3D printed with black PLA, and painted matte black. An inner optical stop of the baffle limits the field of view of the sensor to 60 degrees, while an outer stop prevents direct sunlight from illuminating the inner stop at solar elevations less than 55 degrees. The latter limit was chosen to match the maximum solar elevation at the geographical location of the study.

Two identical sensor systems were constructed and mounted on the drone, one on top of it to measure the sky color and another below the fuselage on the undercarriage to measure the cloak color.

2.3 *The Control System*

To adjust the cloak color to match the sky color, an embedded microcontroller system was built using an MBED LPC1768 module. The module is based on a 96 MHz ARM Cortex M3 microcontroller and has all the necessary infrastructures (GPIO, voltage regulation, buses, etc.) required in this project. In addition, a LoRa radio link was included to provide the capability to manually control and tune the cloak parameters in flight.

Initially, a basic software PID controller was created to control the color matching but was soon found to be unsatisfactory. The integrating I term is necessary to compensate for unknown changes in the LED cloak caused by, e.g., temperature or operating voltage. However, due to the vast dynamic range of the system, I term values high enough to yield a reasonable response time resulted in oscillation under low light levels near the obvious system discontinuity of zero, i.e., darkness. This was remedied by using the PID controller to control the logarithm of the individual color brightness levels instead of the level values themselves. Thus the adjustments *relative* to the brightness level were performed with a constant response time, and the darkness did not present a discontinuity.

The color of the cloak was set with three simultaneous methods. To maintain the best possible color resolution, the 5-bit overall brightness of each single RGB LED was set to the lowest value still sufficient, and the 8-bit R, G, and B values are scaled accordingly. The error between the required and the resulting LED colors was accumulated over the array of LEDs, always carried over to the next LED, performing automatic dithering over the cloak. This was obvious under the darkest circumstances causing most of the LEDs to remain completely black.

The LoRa radio link was realized with an RFM95 868 MHz module. From the ground station, requests were transmitted to the cloak system and status responses were returned back with a rate of 50 Hz. The requests contained commands to turn the cloak on or off and to control the color either purely autonomously or with manual adjustment factors for R, G, and B. The responses reported the measured colors, the calculated LED brightness, and the radio link quality. The brightness information was of interest to verify the margin from the brightest sky color to the cloak maximum output.

The power for the cloaking system was obtained from switching regulators and batteries built into the control system. In drone use, it would be preferential to divert this power from the drone itself.

3 Results

The cloaking system was first applied to a small handheld test patch and then to a commercial drone. Both systems were tested under various natural lighting conditions including the brightest clear and overcast skies. It was found that the cloak constructed with 100% LED strip coverage was indeed bright enough to match the sky, already with ~25% of its maximum brightness. Somewhat unexpectedly, the brightest clear and overcast skies were of almost the same brightness. Since the minimum light output of the cloak is zero, the cloak thus has the capability to match all sky colors, except the solar disk and halo.

The dynamic range of the cloak depends on its size, as the minimum nonzero light output of the entire cloak is the minimum output of a single LED and the maximum is the maximum output of the entire array. Therefore larger cloaks have larger dynamic ranges. The minimum size cloak consists of a single LED and has a dynamic range of 1 : 8191.

3.1 The Test Patch

A 10 cm × 12.5 cm test patch of the cloak was built onto a piece of clear acrylic plate (Fig. 2a). A white patch was placed next to the cloak as a reference. The control system and the sky-measuring sensor were placed behind (on top of) the cloak, and the cloak-measuring sensor was held manually in a position to measure the cloak color. This positioning is relatively noncritical as if the sensor observes only partly the cloak and partly the sky, the control system will still force these to match, i.e., the end result remains the same. The cloak LED array consisted of 144 LEDs and

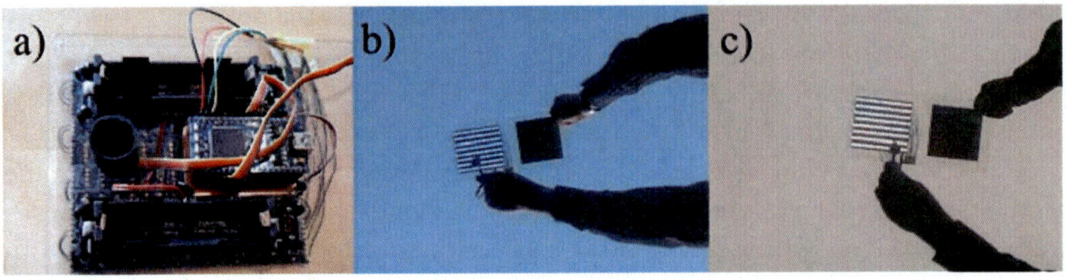

Fig. 2 (**a**) The test patch, (**b**) cloaking against blue sky, (**c**) cloaking against overcast sky

thus covered a dynamic range of 1 : 1 179 648. To power the LEDs, two switching regulators with 5 A output were connected in parallel and driven from their internal 18650 Li-ion batteries.

In Fig. 2b and c, the individual LEDs are clearly visible, but the overall brightness and color of the cloak match the background. This was verified by swapping the two sensors and having the same result. The white reference patch appears as black revealing the huge contrast between the sky and the diffuse ground reflection.

3.2 *The Drone*

For the drone to serve as the carrier for the cloak, we chose a very popular commercial model, the DJI Phantom 4. This drone is of medium size (350 mm diagonal, 1.4 kg), and it is very easy to fly due to its high level of autonomy and advanced collision avoidance features. From previous experience we knew the drone to be capable of lifting at least 1 kg of extra payload. The drone has integrated signaling LEDs at the ends of its motor booms, and these cannot be completely disabled through the user interface. Simply covering the LEDs was unsuccessful as they are very bright and the white plastic hull of the drone is slightly translucent, scattering enough light to make the drone highly visible under low light conditions. So these LEDs had to be removed altogether, which required a complete disassembly of the drone.

The control system and the sky-measuring sensor were mounted on top of the drone, and the cloak-measuring sensor was mounted on the drone landing gear. First flight tests revealed a significant decrease in the GPS visibility for the drone, which was traced back to the close proximity of the GPS receiver antenna of the drone to the control system electronics and power bank of the cloak. The problem was solved by lifting the control system on top of a lightweight supporting platform (Fig. 3). The greater number of LEDs required the power system to be upgraded to 4 parallel 5 A switching regulators powered from a single LiPo battery. The power could not be obtained from the drone itself, as the drone self-diagnostics interpreted it as a power system malfunction.

Fig. 3 The drone equipped with the cloaking system

The underside fuselage of the drone was covered with the LED strip to the best of our ability. The surface in question is not flat but curved in all directions, making the task very challenging. While the strip is flexible, it does not stretch, and it bends easily only in the lengthwise direction. Underside the fuselage, the drone has a gimballed camera and a fixed landing gear, neither of which could be cloaked due to their complex geometry. One end of the removable battery extends slightly outside the fuselage and presents the same problem. Additionally, the drone has several types of downward-facing sensors which were left uncovered. The cloak of the drone consisted of 287 LEDs resulting in a dynamic range of 1 : 2 351 104.

The drone was flown under different lighting conditions to verify the cloaking function (Fig. 4). Both the color and the brightness adapted as expected, and the cloaked parts of the drone disappeared visually after reaching the distance where individual LEDs could not be discerned any longer. The undercarriage, the camera, and the battery end remained visible nonetheless. With a bright background, the propeller disks were slightly visible as dark halos as well. During the darkest test conditions, the dithering function of the cloak became apparent, as most of the LEDs were completely black. While this is evident from the camera shot (Fig. 4c), it was exceedingly difficult to see by the naked eye due to the depth of darkness. The few LEDs that are still emitting light are all at their lowest brightness setting. The ratio of actual brightness between Fig. 4a and c is close to 500 000 : 1.

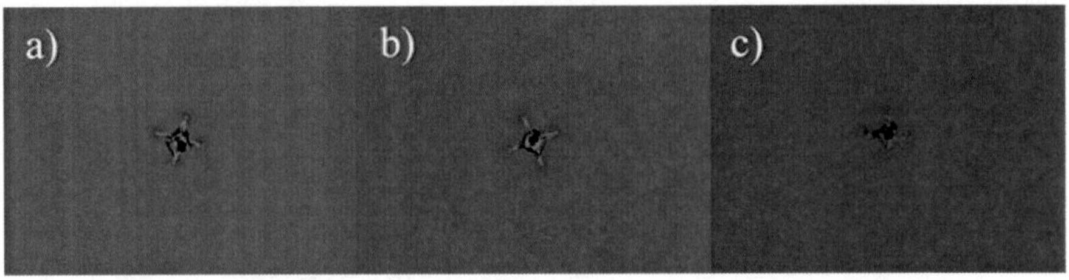

Fig. 4 The cloaked drone under extremely different lighting conditions. (**a**) Bright sky, (**b**) civil twilight, (**c**) nautical twilight

3.3 The Cloaking Factor

The visual acquisition of a drone is a multifaceted process including visual, temporal, environmental, and psychological aspects, and it is therefore quantifiable only through user studies. One of the technical aspects of the cloaking system is the reduction of the visual cross-section it provides to the drone. We measured this from the photographs of the cloaked vs uncloaked drone against different background sky conditions. The sky around the drone in these photographs was verified to be homogenous and thus provided a credible estimate of the drone background (which being occluded by the drone was not observable). The relative cross-section was calculated with MATLAB as the difference of pixel colors between the cropped drone image and a similar size background sample (around the drone) relative to the background size and brightness:

$$\sigma_{DRONE} = \frac{\left|\sum \text{pixels}_{DRONE} - \sum \text{pixels}_{BG}\right|}{\sum \text{pixels}_{BG}}$$

This definition is insensitive toward changes in drone distance and camera parameters, as long as the camera transfer function is linear. Using the difference of the sums instead of the sum of the differences as the nominator allows for the dithering to succeed and also makes the definition insensitive toward camera focus.

The cloaking factor was defined as the ratio of the relative cross-sections of the uncloaked and cloaked drone images:

$$F = \frac{\sigma_{UNCLOAKED}}{\sigma_{CLOAKED}}$$

Thus a cloaking factor F of 1 signifies no cloaking at all, and a cloaking factor of 2 signifies that the remaining relative cross-section is one half of the uncloaked drone. Since the absolute visual cross-section is inversely related to the square of the viewing distance, the acquisition distance d decreases as the cloaking factor increases:

Table 1 Measured cloak effectiveness (including non-cloaked parts)

Lighting condition	Typ. sky luminance (cd/m^2)	Cloaking factor	Relative minimum acquisition distance
Daylight	10 000	3.1	0.57
Civil twilight	1	3.2	0.6
Nautical twilight	0.005	1.6	0.8

$$d_{CLOAKED} = \frac{d_{UNCLOAKED}}{\sqrt{F}}$$

Table 1 summarizes the results of the cloaking experiments. It can be seen that the cloak reduces the visual cross-section significantly. From the photographs in Fig. 4, the heavy impact of the uncloaked camera and undercarriage to these results are clear, but no effort was made to exclude these from the measurements or results.

4 Conclusions

A cloaking system for drones that matches their color to the background sky was constructed and tested. A commercial LED strip was used to build the cloak and proved to be sufficient for all lighting conditions already at 25% of its maximum output. Considering the huge dynamic range necessary for this application, that margin was quite narrow. Nevertheless, at these same conditions, the LED density could be slightly reduced without diminishing performance.

The cloaking system reduces the visibility of the drone down to a fraction, even while large parts of the drone were uncloaked. This allows the drone to operate at a closer distance undetected, which was the original goal.

The commercial drone we used was not an optimal object for cloaking due to the built-in lights, the nonplanar underside, and the external payload. To fully cloak a drone with this approach, a suitable drone should be designed with the cloak ability in mind from the start. Extrapolating from the successfully cloaked parts of our test drone, such a drone would be completely invisible for observers at a suitable distance.

References

1. Commission Implementing Regulation (EU) 2019/947 of 24 May 2019 on the rules and procedures for the operation of unmanned aircraft. http://data.europa.eu/eli/reg_impl/2019/947/oj. Accessed 28 Dec 2020.
2. Federal Regulations (US) Title 14: Aeronautics and space PART 107 — SMALL UNMANNED AIRCRAFT SYSTEMS. https://www.ecfr.gov/cgi-bin/text-idx?node=pt14.2.107. Accessed 28 Dec 2020.

3. Robley, L. P. (1987). Predator: Special visual effects. *Cinefantastique, 18*(1), 34.
4. Bush, V., Conant, J., et al. (1946). Camouflage of sea-search aircraft. In *Visibility studies and some applications in the field of camouflage* (pp. 225–240). Office of Scientific Research and Development, National Defence Research Committee.
5. Lighting Design and Simulation Knowledgebase. Lighting design glossary. *Luminance*. https://www.schorsch.com/en/kbase/glossary/luminance.html. Accessed 28 Dec 2020.
6. Bowditch, N. (2002). *The American practical navigator* (p. 227). Lighthouse press.
7. American Meteorological Society. Glossary of meteorology. *Nautical twilight*. https://glossary.ametsoc.org/wiki/Nautical_twilight. Accessed 28 Dec 2020.
8. Addressable LED strip. https://www.aliexpress.com/item/32806765132.html. Accessed 29 Dec 2020.
9. APA102C Datasheet. http://cdn.sparkfun.com/datasheets/Components/LED/APA102C.pdf. Accessed 29 Dec 2020.
10. VEML6040 Datasheet. https://www.vishay.com/docs/84276/veml6040.pdf. Accessed 29 Dec 2020.

Long-Term Autonomy in Forest Environment Using Self-Corrective SLAM

Paavo Nevalainen, Parisa Movahedi, Jorge Peña Queralta, Tomi Westerlund, and Jukka Heikkonen

Abstract Vehicles with prolonged autonomous missions have to maintain environment awareness by simultaneous localization and mapping (SLAM). Closed-loop correction used for SLAM consistency maintenance is proposed to be substituted by interpolation in rigid body transformation space in order to systematically reduce the accumulated error over different scales. The computation is divided into an edge-computed lightweight SLAM and iterative corrections in the cloud environment. Tree locations in the forest environment are sent via a potentially limited communication bandwidth. Data from a real forest site is used in the verification of the proposed algorithm. The algorithm adds new iterative closest point (ICP) cases to the initial SLAM and measures the resulting map quality by the mean of the root mean squared error (RMSE) of individual tree clusters. Adding 4% more match cases yields the mean RMSE of 0.15 m on a large site with 180 m odometric distance.

Keywords Odometry · SLAM · Sparse point clouds · Lidar · Laser scanning · Forest localization · Autonomous navigation

1 Introduction

The past decade has seen a rapid evolution of methods and technologies in onboard odometry for autonomous navigation and localization. The state of the art has reached a significant level of maturity in both lidar-based [1] and visual-based odometry, among others [2]. Nonetheless, drift in long-term operation is an inherent problem to methods based only on on-board sensors and data, with the probability of a lost position estimation increasing over time [3]. Most methods address this with loop closure [1, 2], where data is compared with older records if locations

P. Nevalainen (✉) · P. Movahedi · J. P. Queralta · T. Westerlund · J. Heikkonen
Department of Computing, Faculty of Technology, University of Turku, Turku, Finland
e-mail: ptneva@utu.fi; parmov@utu.fi; jopequ@utu.fi; tovewe@utu.fi; jukhei@utu.fi

T. Lipping et al. (eds.), *New Developments and Environmental Applications of Drones*,
https://doi.org/10.1007/978-3-030-77860-6_5

are repeated. In any case, long-term autonomy based only on onboard odometry data still presents significant challenges. In remote and unstructured environments such as forests, typical methods do not always apply, and loop closure can rarely be applied [4]. In these scenarios, onboard processing at the edge without network-based computational offloading has inherent limitations. Challenges arise from the point of view of memory (amount of scan data to be stored for later processing), from the perspective of computational load and latency (amount of data to be used for localization through point cloud matching processes), and in terms of the update rate of the localization process (how often is the relative position computed).

Specifically, this chapter deals with the problem of Simultaneous Localization and Mapping (SLAM) in unstructured forest environments with 3D laser scanners. SLAM algorithms aim at tracking the movement of the laser scanner (odometry) related to its surroundings and creating a composition from individual views, which consist of scanned point clouds (PCs), and scanner positions and orientations. A laser scanner attached to a vehicle provides a spatial input signal which can be a very powerful component in supporting situational awareness, especially when fused with input of other sensors.

A structure from motion (SfM) study [5] divides SLAM methods into two divisions: indirect and direct methods and sparse and dense approaches. One can add four more divisions: probabilistic and non-probabilistic methods, structured and non-structured environments [4], fixed and adaptive frame sampling, and sensor fusion approach [6] versus plain SLAM. Indirect methods rely on early frame-by-frame processing, which produces a set of anchor points. The fixed frame sampling uses every frame or a fixed ratio of frames, whereas adaptive sampling tends to skip a sequence of highly similar frames. The case chosen here uses indirect anchor points from adaptively chosen frames, is non-probabilistic, and is aimed to a forest environment, which is non-structured and has nearly uniformly distributed sparse anchor PC. The sixth categorization is "plain SLAM," since this article considers the quality of the resulting environment map independent of the possible fusion steps. Of the excluded fusion techniques, the most important one is Global Navigation Satellite System (GNSS), and the fusion of SLAM and GNSS can be considered partly as propagating the global position of GNSS to individual trees in a most consistent way.

State of the Art Iterative closest points (ICPs) is the standard baseline method in SLAM. It is the computationally most economic choice in its simplest versions, if the convergence can be guaranteed by the application specifics. If a PC is near-uniformly random, the overlap ratio of visible cones is a good estimation for an outlier ratio γ, which is one of the few tuning parameters used by robust ICPs [7]. The outlier ratio γ limits the ICP matching process to $1 - \gamma$ part of the match pairs and reduces the accumulation of the odometric error. If the matches occur in a geometrically consistent zone between two PCs, the match overlap ratio λ can be estimated by $\lambda \approx 1 - \gamma$.

The strategy of forcing a minimum overlap does not guarantee global convergence, though. A globally optimal method with a proven convergence is Go-ICP [8], which can detect a difference between a local and a global ICP match in the case of near-uniform PC. Other "global" methods try to smooth the mean matching error, which is the target function, by various means, but fail to guarantee the convergence to a global optimum.

Near-uniform randomness gives a chance to tighten the conditions of branch-and-bound (BnB) limit estimates used in [8]. We use Go-ICP as a backbone of a naive and risky SLAM method *pcregistericp*() [9–11], since there are no SLAM methods specifically suited for sparse uniformly random PCs to the best of our knowledge.

When dealing with a limited view cone (less than 360° view), a problem similar to closed-loop detection [12, 13] occurs each time the view cone coincides with a much older frame. This can happen in a small scale of 2–5 m, but it can also happen over distances of 0.5 km to 1 km.

Motivation Autonomous mobile robots and specifically unmanned aerial vehicles (UAVs) have seen an increase penetration for forest surveying and remote sensing [14, 15]. Owing to the unstructured environments that forests represent, autonomous navigation presents inherent challenges. Key issues appear in the areas of localization and mapping, where one has to take into account several key points in a local scope to make the SLAM computationally feasible [4]. Moreover, an autonomy stack for forest navigation ought to consider long-term autonomy (e.g., owing to the long distances that UAVs can traverse over long times, or the longer time that ground vehicles can operate). Typical odometry techniques relying on loop closure do not suffice because locations are not repeated often. In particular, we are interested in lidar-based odometry, localization, and mapping with methods that can be used for both unmanned ground vehicles (UGVs) and UAVs.

Taking into account these considerations, there is a need for more advanced techniques for long-term autonomy exploiting registrations of the same objects even from distant locations. This approach differs from traditional loop closure as there can be several partial overlaps of frames over different time scales, and the partial overlap may occur over large distances. Modern laser scanners, even low-cost solid-state LiDARs, are capable of measuring distance to objects up to several hundreds of meters [16].

In addition to the accuracy of localization over long distances, the majority of the state-of-the-art lidar-based SLAM algorithms require relatively high computational resources to operate in real time [1, 17]. Moreover, the amount of points in a single scan PC has increased to millions per second in recent years. Solid-state lidars with limited field of view (FoV) are only able to detect a reduced number of features in a single scan, but the scan density increases significantly. Therefore, techniques that need to compare all points (e.g., ICP) or traditional feature extraction techniques do not scale well. From the perspective of map-based localization, a similar issue arises with approaches such as the normal distribution transform (NDT) [18].

Our approach extends the idea of loop closure to track features (i.e., tree stems) over long-term autonomous operation, which can be stored and compared in the form of sparse PCs. Relying in sparse PCs that contain only uniformly distributed features, we are able to reduce both the processing time for localization (in terms of point cloud matching) and the size in memory of larger-scale maps.

In general, we see a gap in the literature in approaches to long-term autonomy and self-corrective localization leveraging the matching of uniformly random feature points. To the best of our knowledge, this together with exploiting sparse PCs for faster processing in unstructured environments has not been addressed yet. Moreover, our approach can be leveraged for managing and accounting for the actively rotating view cone of modern solid-state lidars.

Recently, there has been progress in rigid body interpolation [19] mainly applied in the robotics field. There is an advantage in having both rotation and translation addressed at the same time. To our understanding, this provides a chance to address the odometric consistency independently of the scale of the localization problem.

Contribution
We propose a method to reduce the cumulative match error by adding extra ICP matches, which comprise a large time interval. Frames within the interval are squeezed together by an interpolation scheme reducing the imprint of sets of anchor points associated together in the final map. The process reduces the noise and blur in the final map, increases the odometric accuracy, and solves both small-scale closed-loop occurrences due to the work cycle movements and large-scale closed-loop problems.

In summary, in this chapter, we address the following three issues: (i) a self-corrective localization algorithm able to incrementally increase the accuracy of the produced environment map without relying on loop closure; (ii) memory efficiency and computation at the edge by relying on sparse point clouds and long-term tracking of features; and (iii) an adaptive approach that adjusts the positioning update rate based on the available data at a given time.

2 Methods

The Site and the Data The test data is from a forest operation site in Pankakangas at Lieksa, Eastern Finland (63°19.08′ N, 30°11.57′ E). The data was recorded in August 2017 in co-operation with participants enlisted in the Acknowledgements section. The sample has a strip road of length 130 m. There are 9009 data frames and 3.7 GB of *.pcap* data. The total number of trees is 680 and one frame includes 130 trees on the average. An example scanner view is depicted in the left detail of Fig. 1. The mean distance to the nearest trees is $L_0 = 3.5$ m excluding the peripheral zone at a distance of 40 m. The right detail shows how the mean distance increases over the radial distance.

Methodology A simple and fast ICP method *pcregistericp*() [10, 11] implemented in Matlab [9] produces rigid body transformations, which can be used to build an environment map from sparse key points, each key point representing a detected

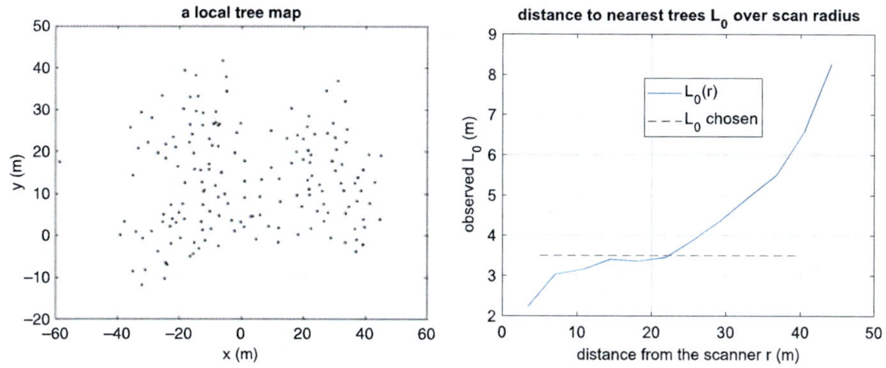

Fig. 1 Left: a single scanner view. Each point represents an edge-computed registration of a tree. Points are 3D even shown in a horizontal projection. Right: mean distances to the nearest trees at each scanner range. $L_0 = 3.5$ m is over all the trees and is depicted with a dash line

tree in a scanner view. The problem is to improve this rather low-quality tree map by selecting a small set of promising pairs of frames and producing a computationally more expensive and more accurate match using Go-ICP [8]. Each extra match between frames j and i may engulf several frames, which should be properly adapted to the newly introduced and very reliable match. The exponential interpolation of rigid body transformations is used for that purpose.

We introduce first the rigid body transformation as a homogenous operator and its logarithm and exponentiation, which are the core of the interpolation method. A novel aspect is the operator power being in the matrix form. We show that the interpolation is contractive meaning that the final PC map improves (individual tree clusters become sharper) per each addition of extra matches. The sharpening can be measured internally by minimization of the mean match error and externally by observing the mean radius of clusters in the final map. Finally, we define the control parameters for the branch and bound (BnB) of the global minimum search of the SLAM match.

2.1 Operator Exponentiation

A rigid body transformation $\tau \in SE(3)$ in the special Euclidean group $SE(3)$ consists of one rotation represented by a rotation $R \in SO(3)$ within the special orthogonal group $SO(3)$ followed by and a translation $p \in \mathbb{R}^3$. The treatise uses the notation and conventions of [19]. The transformation τ can be seen as a homogenous mapping $\tau : q \mapsto q'$:

$$\begin{pmatrix} q' \\ 1 \end{pmatrix} = \overbrace{\begin{pmatrix} [R] & p \\ 0 & 1 \end{pmatrix}}^{[\tau]} \begin{pmatrix} q \\ 1 \end{pmatrix}, \tag{1}$$

where braces [.] depict the matrix representation of an operator. Transformation $\tau(R, p)$ is defined by a pair of a rotation and a translation. An alternative parameterization is $\tau(\vec{\omega}, \theta, v)$, where a unit axis $\vec{\omega} = (\omega_1\, \omega_2\, \omega_3) \in \mathbb{R}^3$ is the rotation axis, θ is a rotation angle around that axis, and $v \in \mathbb{R}^3$ is the tangential direction, where the origo moves in the beginning of that rotation. Note that $\vec{\omega}$ is the unit eigenvector of $[R]$ associated with the eigenvalue 1: $[R]\vec{\omega} = \vec{\omega}$.

A twist $S([\omega], v)$ combines two of the elements of a rigid body transformation, and its matrix form is

$$[S] = \begin{pmatrix} [\omega] & v \\ 0 & 0 \end{pmatrix}, \tag{2}$$

where $[\omega]a = \vec{\omega} \times a$ is a vector cross product for any $a \in \mathbb{R}^3$, or, as written open in the matrix form:

$$[\omega] = \begin{pmatrix} 0 & -\omega_3 & \omega_2 \\ \omega_3 & 0 & -\omega_1 \\ -\omega_2 & \omega_1 & 0 \end{pmatrix}. \tag{3}$$

Note that this definition gives $[\omega]$ a cyclic property: $[\omega]^3 = -[\omega]$, which will be used, when dealing with series expansions of e^x, $\sin x$ and $\cos x$. Exponentiation of $[S]\theta$ gives us

$$e^{[S]\theta} = \begin{pmatrix} e^{[\omega]\theta} & G(\theta)v \\ 0 & 1 \end{pmatrix} = [\tau], \tag{4}$$

where the right equality can be settled by setting $v = G^{-1}(\theta)p$ and $e^{[\omega]\theta} = [R]$. The twist gain function $G(\theta)$ unfolds by the exponentiation series and the rotation matrix term can be expanded to a closed form:

$$[R] = I + \sin\theta[\omega] + (1 - \cos\theta)[\omega]^2, \tag{5}$$

which is the well-known Rodriguez formula. Finally, raising τ to a power $u \in \mathbb{R}$, one gets

$$[\tau^u] = \begin{pmatrix} [R^u] & p_u \\ 0 & 1 \end{pmatrix}, \tag{6}$$

where

$$[R^u] = I + \sin\theta u[\omega] + (1 - \cos\theta u)[\omega]^2 \tag{7}$$

$$p_u = G(\theta u)G^{-1}(\theta)p. \tag{8}$$

The homogenous representation allowed a definition of a matrix power of a rigid body transformation limited to $SE(3)$. To signify this limitation, we write $[\tau^u]$ and not $[\tau]^u$, since the wide realm of general matrices is perilous [20], what comes to exponentiation and taking logarithms. The same argument holds to the notation with $[R^u]$.

The twist gain function $G(\theta)$ is opened next:

$$G(\theta) = I\theta + (1 - \cos\theta)[\omega] + (\theta - \sin\theta)[\omega]^2. \tag{9}$$

Its inverse is needed in Eq. 8:

$$G^{-1}(\theta) = \frac{1}{\theta}I - \frac{1}{2}[\omega] + \left(\frac{1}{\theta} - \frac{1}{2\tan(\theta/2)}\right)[\omega]^2. \tag{10}$$

One can easily see that there is a singularity in $G^{-1}(\theta)$ when $\theta \to 0$. But the product in Eq. 8 stays defined, albeit it needs a Taylor series expansion[1] at $\theta = 0$. This is needed because the homogenous formulation chosen here is not a conformal theory [21]. The product $G(\theta u)G^{-1}(\theta)$ develops to

$$\begin{aligned} G(\theta u)G^{-1}(\theta) = Iu &+ \left[A(\theta, u) - \frac{\sin\theta u}{2}\right][\omega] \\ &+ \left[u - \frac{1 - \cos\theta u}{2} - B(\theta, u)\right][\omega]^2, \end{aligned} \tag{11}$$

where

$$A(\theta, u) = \frac{1 - \cos\theta u}{2\tan\theta/2}, \quad B(\theta, u) = \frac{\sin\theta u}{2\tan\theta/2}, \tag{12}$$

which are both of a form 0/0 at $\theta = 0$. Taylor series developed at $\theta = 0$ gives

$$A(\theta, u) = \tfrac{u^2}{2}\theta - (u^2 + u^4)\theta^3/24 + \mathcal{O}(\theta^4) \tag{13}$$

$$B(\theta, u) = u - (u/12 + u^3/6)\theta^2 + \mathcal{O}(\theta^4). \tag{14}$$

[1] ...or a min–max polynomial definition, which is excluded from this treatment.

Note that small values of θ will often occur with the intended application, whereas large values $\theta \approx \pi$ occur seldomly, if ever.

Extraction of $[\omega]$ and θ from a given $[R]$ is called taking a rotation logarithm, since $[R] = e^{[\omega]\theta}$. The exact logarithm algorithm is given in [19] and has two special cases for $\theta \approx 0$ and $\theta \approx \pi$. The intended application of the transformation matrix power is such that one needs to solve Eq. 6 several times with different values of u, so the constant parts $[\omega]$ and θ are pivotal. Figure 2 shows an example, where a rigid body coordinate frame τ_a is interpolated to another frame τ_b using 11 values $u \in [0, 1]\mathbb{R}$.

The matrix power in Eq. 6 has one special case of pure translation where there is no rotation ($\theta = 0$ and $[R] = I$) with

$$\begin{pmatrix} I & p \\ 0 & 1 \end{pmatrix}^u = \begin{pmatrix} I & pu \\ 0 & 1 \end{pmatrix}. \tag{15}$$

Naturally, this special case should be covered by a general solution of the vector p_u. As a sanity check, setting $\theta \approx 0$ leads to $p_u \approx pu$ for all $u \in \mathbb{R}$. By setting $u = 1$ and after a tedious trigonometric manipulation, one gets $p_u = p$ for all $\theta \in [0, \pi)$. Although Eqs. 11–14 are novel in the context of the matrix power of the homogenous formulation, a similar Taylor series approach has been presented for dual quaternion exponentiation and logarithm in [22], and one could construct a similar transformation power τ^u using suitable quaternion libraries. The value of Eqs. 11–14 is that the odometric process described in the next section can proceed within a usual matrix infrastructure. The computational price tag of two alternative formulations (dual quaternions and homogenous co-ordinates) for taking multiple matrix powers is surprisingly close to each other for large PCs.

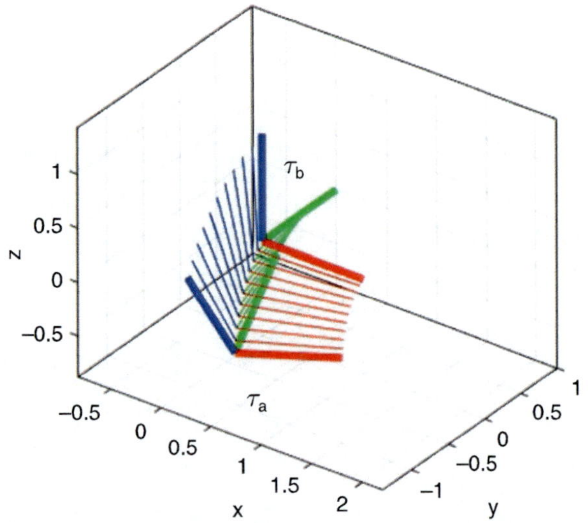

Fig. 2 Example of a power interpolation $\tau_u = \tau_a^{1-u}\tau_b^u$. When $u = 0 \rightarrow 1$, $\tau_u = \tau_a \rightarrow \tau_b$ smoothly between two rigid body frames τ_a and τ_b. Columns of $[R]$ form the orthogonal frame axes, which are shown in red, green, and blue

A study on the error of the exponentiation of the rigid body transformation follows. Two consecutive operations $\|[(t^u)^{1/u}] - [t]\|$ produce a Frobenus norm error shown in Fig. 3. The result is the average over 300 rigid body transformations with a uniformly distributed $p \in [-1, 1]^3 \subset \mathbb{R}^3$ and $R(\vec{\omega}, \theta)$, where $\vec{\omega} \in S^2$ is uniformly distributed over a unit sphere $S^2 \subset \mathbb{R}^3$ and $\theta \in [0, \pi)$ is also uniformly distributed. The sequential matrix multiplication version with $1/u \in \mathbb{N}$ has been provided (red line) alongside the usual $u \in [0, 1] \subset \mathbb{R}$ matrix power test (blue line). As can be seen from Fig. 3, the computational accuracy is not a problem near $u = 0$, and $u \approx 1$ does not usually occur. The overall accuracy in the multiplication case is 4.8×10^{-3}, which is enough for practical implementation. Both tests are clearly conservative when compared to actual computations, which generate new operators τ^u from constant values θ and $[\omega]$.

2.2 *Rigid Body Motion Interpolation*

This section expands the presentation in [23] and uses the notation of [19]. The detailed definitions are provided since the formulations come from a variety of sources. Odometry is built by matching sequential PCs P_l and P_{l+1} in the coordinate system of the frame l by estimating a transformation $\tau_{l+1\,l}$ from a frame $l+1$ to the frame l. A simple and fast ICP method *pcregistericp*() of Matlab [9–11] is applied

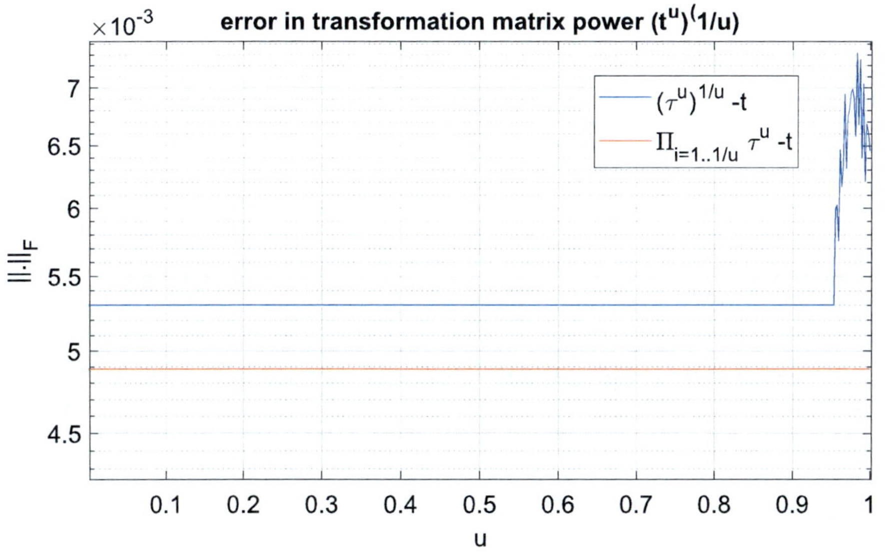

Fig. 3 A numerical verification of rising rigid body transformations to a power u. Blue line depicts taking a power u and then $1/u$. Red line is for taking a power u followed by a number $(1/u \in \mathbb{N})$ of matrix multiplications

to produce a sequence of rigid body transformations $T = \{\tau_{l+1\,l} \mid 1 \leq l \leq n\}$ from frames $l + 1$ to frames l. This process is not secure, and it is possible to have an erroneous match, which is off some 2–10 m.

The combination of two PCs achieved by a successful match is denoted by $.\tilde{\cup}.$ as $P_l \tilde{\cup} P_{l+1} t_{l+1\,l}$, where the transformation matrix $t_{ij} = [\tau_{ij}]^T$ because, unlike in the definition of Eq. 1, points are now columns of a PC matrix. A match between frames $l + 1$ and l includes inaccuracies e_l and identification of outliers (points not matched to any point) and of matching pairs of points. The final SLAM result has all frames matched to the first frame:

$$P = \tilde{\cup}_{l=1}^{n} P_l\, t_{l1}, \tag{16}$$

where n is the number of frames, $t_{11} = I$, and the total transformation matrices t_{l1} are built iteratively from local matches $t_{l+1\,l}$ by $t_{l+1\,1} = t_{l1} t_{l+1\,l}$.

The PCs $P_l\, t_{l1}$ each contribute to the total map P. A typical ICP process produces a chain of stepwise transformations T. One can recover a random transformation connecting frames i and j by

$$t_{ij} = t_{j1}{}^{-1} t_{i1}. \tag{17}$$

An application of Eq. 16 is called globalization (or SLAM process). The odometry problem is solved when the globalized translation vectors are extracted. The path Q of the vehicle is

$$Q = \{q_{i1} \mid \tau_{i1} = (R_{i1}, q_{i1})\}_{i=1\ldots n} \tag{18}$$

The basic scenario of the self-corrective odometry is depicted in Fig. 4 using two paths Q and Q' to represent a situation, where some of the transformations $t_{l+1\,l},\ j \leq l < i$ have been judged inaccurate, noisy, or inexact by some criteria. The criteria is usually related to the blurriness of the global map. Then, a corrective check is being performed from the frame i to the frame j producing an improvement of a match. Formally, an error measure $e(P_j \tilde{\cup} P_i\, t_{ij}{}') < e(P_j \tilde{\cup} P_i\, t_{ij})$ of a match improves, when a new match $t_{ij}{}'$ is used instead of a synthesized transformation t_{ij} of Eq. 17. Now, all the intermediary PCs $PC_l,\ j < l < i$, need to be updated.

Let $\Delta\tau_l$ be correction needed at a frame l: $\tau_{lj}\Delta\tau_l = \tau'_{lj}$, and let it be forced to have boundary conditions $\Delta t_j = I$ and $\Delta t_i = \Delta t = t_{ij}{}^{-1} t_{ij}{}'$. A rather obvious interpolator is by $\Delta t_l = [\Delta\tau^{u_l}]^T$, where $0 \leq u_l \leq 1$ with obvious end conditions $u_j = 0$ and $u_i = 1$. Assuming representative powers u_l are defined for each transformation t_{lj}, one can solve new values $t_{lj}{}'$:

$$t_{l1} := t_{l1}{}' = t_{j1} t_{lj} [\Delta\tau^{u_l}]^T,\ j \leq l \leq i, \tag{19}$$

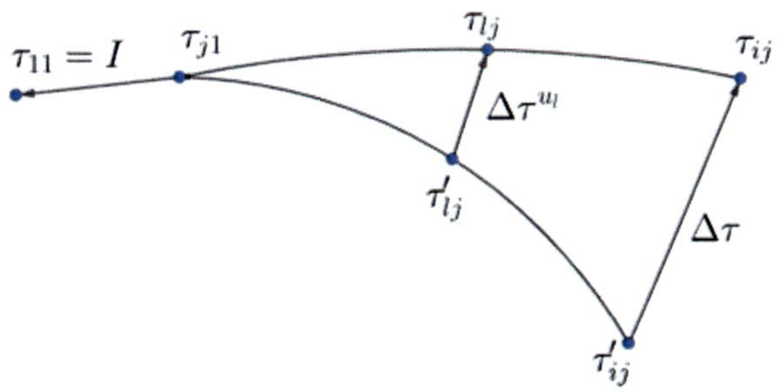

Fig. 4 An update from τ_{i1} to τ_{i1}' makes the old path $j+1, \ldots, i-1$ incompatible. It can be corrected by any interpolation scheme, e.g., Eq. 19

where ":=" denotes a computational substitution of a new value. As a sanity test, by setting $l = j$ and $u_l = 0$, one gets $t_{j1}' = t_{j1}I^2 = t_{j1}$. And by setting $l = i$ and $u_l = 1$, one gets
$t_{i1}' = t_{j1}t_{ij}t_{ij}^{-1}t_{ij}' = t_{j1}t_{ij}'$. The path following after the frame i changes after this update as well. The rest of the frames have to be corrected to align properly with the updated value t_{i1}':

$$t_{k1} := t_{i1}' \overbrace{t_{i1}^{-1} t_{k1}}^{t_{ki}}, \; i < k \leq n \,. \tag{20}$$

One question remains: how to choose the power u_l, given a SLAM history $\{t_{l1}\}_{j \leq l \leq i}$? There are several possibilities but for numerical experiments we used the simplest possible strategy, the relative continuous index:

$$u_l = (l - j)/(i - j), \; j \leq l \leq i. \tag{21}$$

Contractive Property We propagate change on odometric sub-path $j \leq l \leq i$ by using $[\Delta\tau^{u_l}]$ as a correction term. As long as all the involved powers u_l are confined to the unit interval $0 \leq u_l \leq 1$, the new PC map $\overset{\sim i}{\cup}_{l=1} P_l t_{l1}'$ contracts, i.e., all the involved rotations θ_{lj}' for each sub-match $\tau_{lj} = \tau([\omega_{lj}], \theta_{lj}, q_{lj})$ of a corrective step between frames i and j become smaller $\theta_{lj}' \leq \theta_{lj}$ and the magnitude of translations q_{lj}' gets reduced $q_{lj}' \leq q_{lj}$. The proof is based on the monotonicity of terms $G(\theta u)G^{-1}(\theta)$ (see Eq. 8) and θu in the base $\{I, [\omega], [\omega]^2\}$. A visual evidence of this is shown in [23], where tree clusters get less dispersed on each step of iterative improvement.

2.3 *Branch-and-Bound Limits*

The globally convergent ICP method Go-ICP [8] uses two coefficients σ_t and σ_r to set up the granularity of the transition and rotation search space in the BnB search grid, respectively. Two coefficients $\gamma_t = \sqrt{3}\sigma_t$ and $\gamma_{rp} =$

$2\sin(\min(\sqrt{3}\sigma_r/2, \pi/2))\|p\|$ define the local lower bound of the minimum match error at a point $p \in P$ in the original scanning frame. Note that rotational term γ_{rp} indeed depends on the point p. A minimum bound $\underline{e}_{ij}$ of a match error e_{ij} between frames i and j is

$$\underline{e}_{ij}^2 = \sum_{p \in P} \left(\max(e_p - \gamma_{rp} - \gamma_t, 0)\right)^2. \tag{22}$$

Tree locations in a forest usually have a nearly uniform distribution, which can be described by a mean distance L_0 m between natural neighbors (a concept defined in the next paragraph). The right detail of Fig. 1 depicts the L_0 distribution over the scanning range r, and even the point density depends on range, the zone $10 \leq r \leq 26$ m with $3.0 \leq L_0 \leq 4.0$ m is large and populated enough for our purposes. In our data samples, $L_0 = 3.5$ m was the observed average specific to the data collection site.

To make the definition of L_0 more formal, we define natural neighbors $q \in N(p) \subset P$ of a point p as those points, which get connected by an edge $(p, q) \in E \subset P^2$ in a Delaunay triangularization (P, E, T) of a point set P. There, E are edges of Delaunay triangles $T \subset P^3$. Then, the mean distance is

$$L_0 = \underset{p \in P}{\text{mean}} \underset{q \in N(p)}{\text{mean}} \|p - q\|. \tag{23}$$

If a magnitude $\delta = \|q\|$ of a pure translation from a perfect match is smaller than $L_0/2$, $\delta < L_0/2$, the ICP convergence is very likely. This will be shown later by a numerical experiment. We define this limit as δ_0:

$$\delta_0 = L_0/2. \tag{24}$$

For reference, a hexagonal lattice is the optimal packing on points and having two such PCs switched randomly produces a mean match error $e = 0.35L_0$.

Other important parameters characterizing the scanned PCs are the scanning scope R and the allowed outlier ratio γ, which makes the standard ICP method somewhat more robust. Outliers are points without a proper match. Figure 5 shows a circular PC being rotated by an angle θ_0. At a distance r_0, $\delta_0 = r_0\theta_0$. The radius r_0 divides the disc into two parts with ratios $\gamma : 1-\gamma$. A simplifying assumption is being made that all the outer point pairs do not match, and all the inner point pairs do match, so that $1 - \gamma = \pi r_0^2/(\pi R^2)$, and one can solve θ_0:

$$\theta_0 = \frac{\delta_0}{\sqrt{1-\gamma}R}. \tag{25}$$

A justification for this simple derivation of the limit θ_0 is that a rotation with $\theta_0 < \theta$ of two identical uniformly distributed PCs produces a situation similar to a case of two i.i.d. PCs at the outer zone $r_0 < r$. Both a translation δ_0 and a rotation

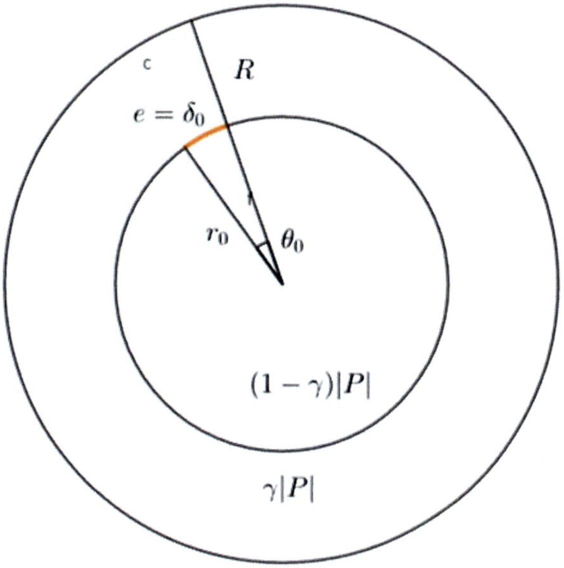

Fig. 5 A rotation θ_0 produces a mismatch δ_0 to $\gamma|P|$ point pairs starting from r_0, when a uniformly distributed PC is rotated around its center

θ_0 are related to the so-called correlation length [24], which is a distance where the auto-correlation of a spatial sample vanishes.

A simple ICP is assumed to succeed when:

$$\left(\frac{\delta}{\delta_0}\right)^2 + \left(\frac{\theta}{\theta_0}\right)^2 \leq 1, \tag{26}$$

where δ is the known magnitude $\delta = \|q\|$ of the known match $\tau([\omega], \theta, q)$ and θ is the known horizontal rotation from the correct match. This means that for a possible grid search or BnB approach, the two parameters λ_0 and θ_0 define the grid granularity.

A complete misalignment has mean error e between pairs of matching points $e \approx \delta_0$, and a complete alignment equals the registration noise ϵ: $e \approx \epsilon$. The ICP match succeeds at the limit $\theta = \theta_0$ since the point density decreases, and the local L_0 increases, when scan radius r grows, see the right detail of Fig. 1.

The BnB search grid can be set to a granularity, where the final attempt at the finest level of the search hierarchy can be safely done by a simple ICP. By this arrangement, the BnB grid does not need to extend to actual tolerances sought after. Numerical values in our example data are $L_0 = 3.5$ m, $\gamma = 0.6$, $\delta_0 = 1.75$ m, and $R = 35$ yielding $\theta_0 = 3.7°$.

The convergence condition of Eq. 26 requires a verification. Figure 6 summarizes a test setting, where matched PC pairs drawn from the data were artificially separated by $\tau(\theta, \delta q^0)$, where translations were taken to several directions encoded by unit vectors q^0 to register the effect over the translation range δ. Two contour lines with the match error $e = 0.2$ m and $e = 0.7$ m are shown. A successful match

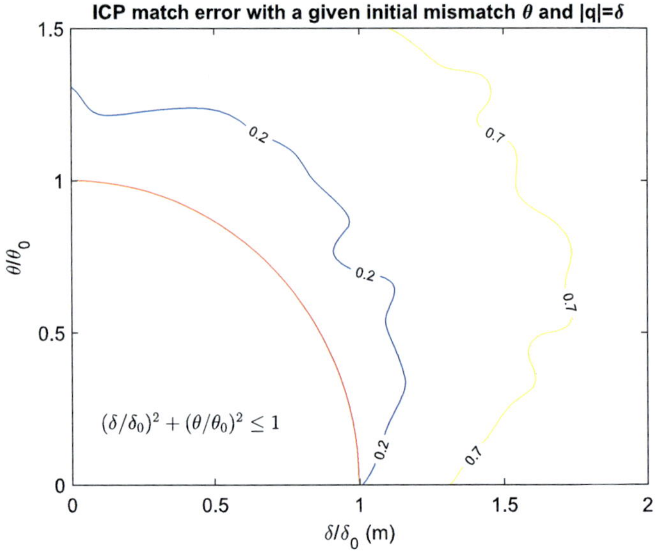

Fig. 6 Match error e (m) of a simple ICP method, when initial mismatch state $\tau(\theta, \delta)$ is known. The area inside the circle is where the convergence is guaranteed for uniform distributed PCs with $L_0 = 2\delta_0$

has $e \leq 0.2$ mm since this does contribute well to the desired final tree map accuracy. The point registration noise ϵ from the tree registration process is approximately $0.05 < \epsilon < 0.1$ m. The convergence area of the condition of Eq. 26 is inside the red arc. 200 PC pairs were used to produce the plot.

The final values for the Go-ICP granularity coefficients are $\sigma_t = \delta_0$ and $\sigma_r = \theta_0$. Each match uses an iteration stop criterion given in [8], and the only extra control layer is by monitoring that two PCs have enough geometric overlap in the matched configuration. The overlap $\lambda \approx 1 - \gamma$, but a geometric calculation using view cone characteristics is used for the actual test. This is because frame PCs contain churn; trees obscure each other and some outliers occur everywhere in the scanning view. If λ is not large enough, $\lambda < \lambda_0 = 0.4$, the frame pair (i, j) will not be used in the iterative improvement of the matches.

The odometry is done in 3D and σ_r concerns also the roll (and pitch) of the vehicle, even these had a negligible effect in point matching. This is because the point cloud is relatively flat, see Fig. 7. The figure depicts also a limit chosen $\phi_0 = 8.2°$ for a successful match. This was found by a numerical test producing a similar plot as shown in Fig. 6. The size of ϕ_0 indicates that the BnB search grid should be elongated (it is cubic grid in Go-ICP implementation). Very large rolls or pitch movements did not occur, and so we limited the BnB search space of rotation to $\pm 30°$ horizontal zone and trusted that the hierarchical BnB quickly eliminates the useless search space. So, the final ICP convergence test is $(\delta/\delta_0)^2 + (\theta/\theta_0)^2 + (\phi/\phi_0)^2 \leq 1$.

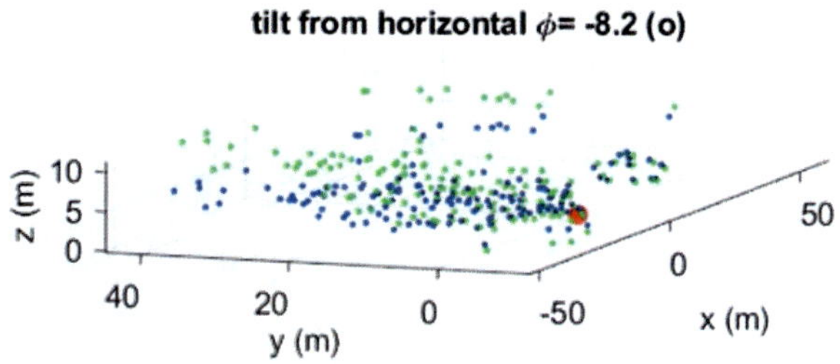

Fig. 7 A demonstration of the found limit case for a rather certain ICP match to occur. The angle $\phi = 8.2°$ is the deviation from the vertical axis

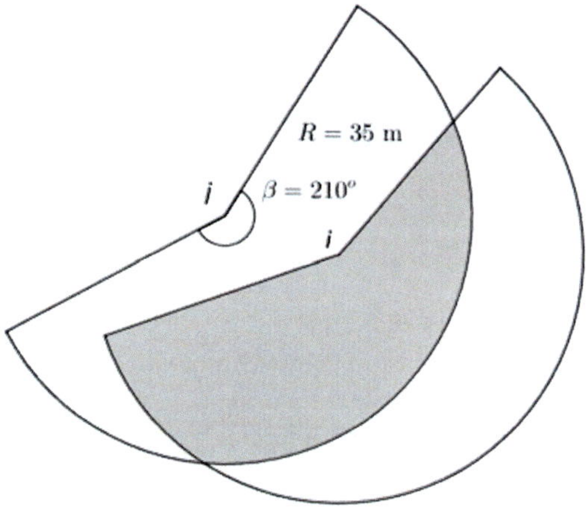

Fig. 8 Overlap (gray area) can be estimated from the orientation and position only under an assumption of uniform PC distribution

2.4 *Iterative Improvement*

An initial tree registration and SLAM over sparse PCs is to be done at the autonomous vehicle. The aim of the initial SLAM is the immediate collision avoidance and basic orientation along the vehicle task goals. The PCs of selected views $V = \{(PC_l, \tau_{l1}\}_{l=1,\dots,n}$ will be sent to the cloud environment, where an improvement of the map will occur.

The estimation of the overlap λ is depicted in Fig. 8. Two vehicle poses τ_{i1} and τ_{j1} in views V are depicted by their view cones. We noticed that the vertical dimension and the corresponding rotations corresponded very little to the final SLAM map via the match pair selection. Therefore, for example, the overlap analysis is done in a projective horizontal plane.

The process starts by selecting m potential pairs of frames, which will be subjected to improvement. Figure 9 depicts a scatter plot based on estimated overlap of views V based on the relative view cone overlap λ and the mean error $e(P_j t_{j1} \tilde{\cup} P_i t_{i1})$ of the match, where non-overlapping parts are excluded. The maximum frame difference max $i - j = 1000$ and 0.2 % of the inspected 1,870,000 frames fall to a promising or acceptable set. The acceptable set was found by looking overlap ratios over $0.2 < \lambda$ and checking the match error $e(P_i t_{i1} \tilde{\cup} P_j t_{j1})$

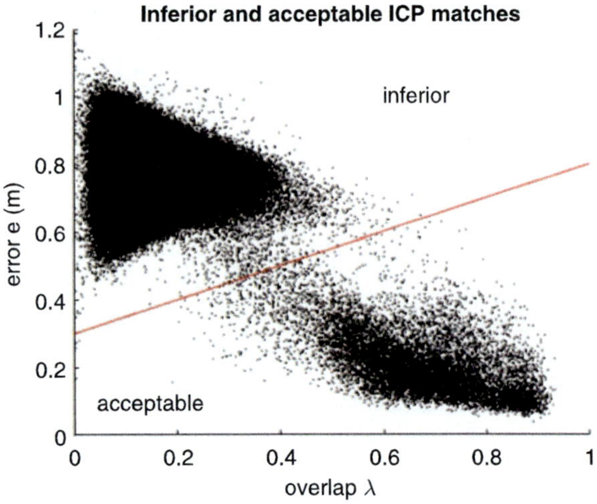

Fig. 9 Scatter plot over ICP matches. An ICP match attempt either succeeds (acceptable) or is inferior (either fails completely or produces small overlap and large error)

of pairs (i, j). Note that value pairs needed (match errors e and geometric overlap factors λ) are fast to achieve in practice, e.g., the overlap factor can be estimated by any universal approximator, e.g., a small specialized convolutional neural network (CNN). In a real-time application, this initial production of values (e, λ) can be done by random tests.

The preliminary set $I_2 \subset [1, n] \subset \mathbb{N}$ depicted by 3500 black dots is sampled to a subset of $m \approx 100$ nearly Poisson disk distributed pairs depicted by red circles in Fig. 10. The detail at the top of Fig. 10 is a schematic about how each black dot $\iota = (i, j) \in I_2$ relates to two frames j and i. The nearly Poisson disc sampling was chosen since one can assume an individual sample will improve the surrounding pairs with equal amount everywhere. A mini-algorithm for producing a promising sample selection I follows:

1. Test recent views (P_l, τ_{l1}) randomly and select pairs (i, j) with $0.2 < \lambda$. From those, select ones with the following condition fulfilled:

$$e_{ij} < 0.3\text{m} + \lambda_{ij} \times 0.5\text{m} \tag{27}$$

   and add $(i, j) \in I_2$. The inequality border is depicted by a red line in Fig. 9. Note that evaluation of values e_{ij} and λ_{ij} is relatively cheap, since the former comes from a direct nearest neighbor search and the latter is estimated directly from the parameters of transformations $\tau_{l1}(\theta_l, \vec{\omega}_l, p_l)$ with $l \in \{i, j\}$.
2. Round I_2 to a set of grid points with spacing ϵ. A set rounding operator $\{.\}_\epsilon$ is introduced for that purpose: $\{A\}_\epsilon = \{\text{round}(a/\epsilon)\epsilon \mid a \in A\}$. The spacing ϵ is decreased iteratively by $\epsilon := 0.8\epsilon$ until the size $|I_1|$ of the rounded set is closest to the intended size: $|I_1| \approx m$. The initial guess is $\epsilon = \sqrt{\text{box area in Fig. } 10/m}$. With the final ϵ,

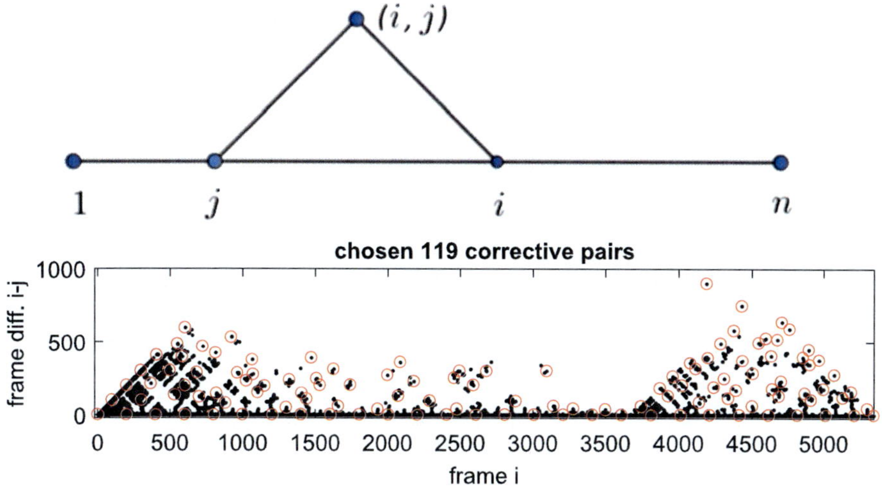

Fig. 10 The selection of extra matches. Above: a match pair (i, j) representing a potential match is related to frames i and j. The original SLAM proceeds along the abscissa from 1 to n with $i = j + 1$. Below: set I (red circles) of improvement matches with nearly Poisson disk distribution is chosen from 2800 potential match pairs of a set I_2 (black dots)

$$I_1 = \{I_2\}_\epsilon. \tag{28}$$

3. For each occupied grid point $\nu \in I_1$, choose the nearest match from the set I_2:

$$I = \{\iota \mid \iota = \underset{\mu \in I_2}{\operatorname{argmin}} \|\mu - \nu\|,\ \nu \in I_1\}. \tag{29}$$

The main step of the algorithm applies Eqs. 19 and 20 randomly until the whole set I is exhausted or a convergence criterion is fulfilled. The simple SLAM is tried first. If it delivers a match error e and an overlap λ which do not fulfill the condition of Eq. 27, Go-ICP is called instead.

2.5 *Quality Criteria of the Final Map*

Basically, there are two possible convergence criteria, one expressing the mean match error e_J over a subset $J \subset I_1$ and another one quantifying the quality of the final map. A measure useful for possible applications of tree maps is the tree registration noise e_C [23]. The registration noise is root mean square error (RMSE) of the tree cluster points from the arithmetic mean of the cluster.

This study focuses on finding the best possible transformations, so we use a numerically faster measure, which addresses the sharpness of the resulting map image. For that purpose, two grid factors $\epsilon_1 = 0.2\,\mathrm{m}$ and $\epsilon_2 = 10.0\,\mathrm{m}$ are chosen. The first one counts $1\ldots4$ grid points for a tree with a diameter $D = 0.1\ldots0.2\,\mathrm{m}$, and the second one is conveniently larger than the mean distance between nearest trees $L_0 = 3.5\,\mathrm{m}$ given in Sect. 2.3. One can define a blur ratio $0 < \beta < 1$:

$$\beta = \frac{|\{P\}_{\epsilon_1}|\epsilon_1^2}{|\{P\}_{\epsilon_2}|\epsilon_2^2}, \tag{30}$$

where $P = \tilde{\cup}_{l=1\ldots n} P_l t_{l1}$ is the SLAM map, and $\{.\}_\epsilon$ is the set rounding operator originally defined for Eq. 28. The numerator of the ratio in Eq. 30 approximates the occupied area in the final map P and the denominator estimates the overall area of the map.

The blur ratio β is used as a target parameter to be minimized in the iterative improvement. It is related to the tree registration noise e_C by having the minimum at the same time, but the absolute value of β depends on how much undergrowth and small trees are on the site. Using β in governing the improvement process is a novel feature, e.g., the authors of [23] use the registration noise e_C instead, which requires alpha shape [25] clustering. The tree clusters are detected by using $r_\alpha = 0.5\,\mathrm{m}$ as the alpha shape radius and ignoring clusters with less than 15 points.

3 Results

The initial (top) and end states (bottom) of the final map $P = \tilde{\cup}_{l=1}^{n} P_l t_{l1}$ are depicted in Fig. 11. Unlike with the ordinary ICP match $.\cup.$, the associations between matching points have not been created, but the map P is just an unstructured PC. The odometric path is plotted in red. The blur ratio β of Eq. 30 moves from the initial $\beta = 0.31$ of the top map of Fig. 11 to $\beta = 0.022$ of the end map at the bottom. A tree map of practical applications usually consists of tree cluster centers only, but for the sake of illustration, the registration points from all frames have been included. The initial scan on the top detail reaches up to 60 m distance, while consistent tree registrations in the below map are within a 40 m stripe. The simplistic SLAM method *pcregistericp*() has a large rotation error, which is seen from the elongated tree clusters in the top details. The final map is the result of the medium-gaps-first strategy. Some blurred details are undergrowth and thickets.

The average stepwise match error $e = \mathrm{mean}_{k=2}^{n}\, e(P_{k-1} \tilde{\cup} P_k t_{k\,k-1})$ was rather stable over the iterative improvement. This is probably because as some tree clusters get sharper during the process, and some spread out due to registration errors. The order of choosing the corrective pairs of frames $\iota = (i, j)$ had a great effect. The strategy of choosing the smallest corrective steps first ($i - j$ in ascending order) ends to a worse end result $\beta = 0.035$ than the medium-gaps-first strategy, which reaches $\beta = 0.022$. This result is depicted in Fig. 11. Three strategies, medium-gaps-first,

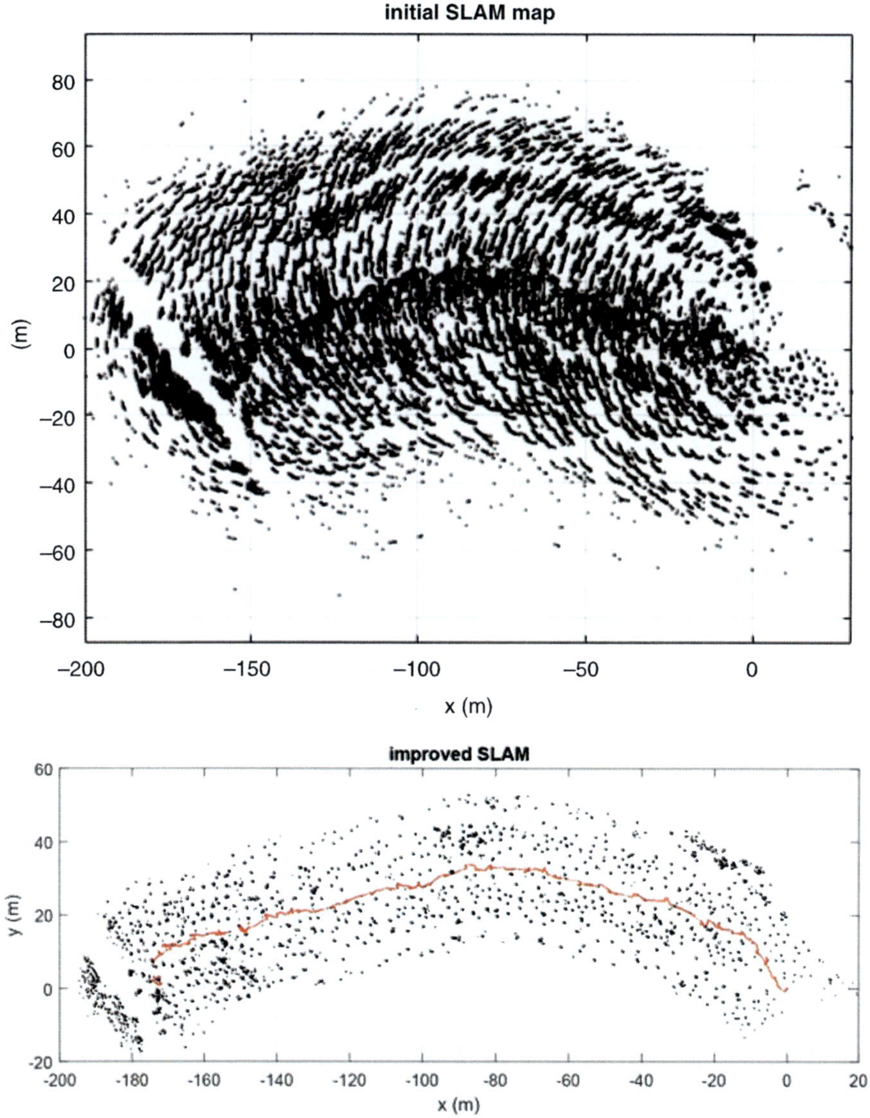

Fig. 11 Above: the initial SLAM map. Below: the final map by the medium-gaps-first strategy

smallest-gaps-first, and random choice are summarized in Table 1. The map noise e_C is the RMSE of the tree cluster radius.

The map noise value found $e_C = 0.15\,\mathrm{m}$ is competitive when compared to a similar study [26] with $e_C \approx 0.2\,\mathrm{m}$ and with more dense original scanning. The test sites of our study and of [26] also compare well: our area was $180 \times 40\,\mathrm{m}^2$ with approximately 600 trees with mean distance $L_0 = 3.5$ m and [26] has a $300\,\mathrm{m} \times 35\,\mathrm{m}$

Table 1 Evaluation times of two ICP methods.

Method	Blur factor (1)	RMSE error (m)	# corrections (m)
Small-gaps-first	0.035	0.27	119
Medium-gaps-first	0.022	0.15	71
Random order	0.051	0.19	119

Table 2 Evaluation times of two ICP methods

Method	One call (s)	Number of calls
Go-ICP	0.41	29
pcregistericp	0.19	80

Table 3 The BnB search space reduction, when Go-ICP gets adapted to the nearly sparse uniform PCs

PC type	Horiz. zone (°)	Rot. granul. σ_r (°)	Transl. granul.σ_t (m)	BnB size
General	180	1.0	1.2	350×10^6
Sparse uniform	60	3.7	1.9	1.5×10^6

area (much narrower) with 224 trees with a mean distance $L_0 = 7.5$ m (much more mature trees). We had two paths whereas [26] had one path.

A Python implementation of Go-ICP [27] was used. The average run time was 0.41 secs over 29 Go-ICP runs triggered. The average size of the PCs was 131 points making the Go-ICP quite fast (Table 2).

The standard Go-ICP [8] uses much smaller coefficients σ_r and σ_t, since it is not specialized to near uniform PCs. Also, we found that the tilt ϕ from the vertical axis of the vehicle is limited to $\phi \in [-\pi/6, \pi/6]$, and this further reduces the BnB search space. These advantages are summarized in Table 3. The translation granularity σ_t is computed in [8] by $\sigma_t = 10^{-4} \times N \times L/2$, where $N \approx 130$ is the data point number and $L \approx 180$ m is the largest included diameter of the PC. This gives an automated value $\sigma_t = 1.2$ m. The translation search was limited to $10 \times 10 \times 2\,\text{m}^3$ volume in both cases.

4 Discussion

Figure 10 shows how the slow movement and constant changes in view orientation of the vehicle at frames 0–500 and 3700–4000 causes a lot of possible correction matches demonstrating potential for frame selection, e.g., by dead-reckoning criteria based on inertial mass units (IMUs) and short-term SLAM. This addition does not change the algorithm proposed, though.

At the frame 3500 of Fig. 10, there is a moment where the vehicle moves directly and fast ahead leaving very little need for iterative SLAM improvement. This phase is close to usual movement of automobiles and is a dominant mode of test cases presented in the existing literature. Our proposal takes into account possible work cycles and constant back and forth translation and irregular rotational movement, e.g., what happens in the forest conditions.

More experiments are needed in deciding a sensible strategy over the application order of improvement matches. The medium-gaps-first strategy is just a best found for this particular task, and obviously there is need for some sort of control, e.g., an end condition to stop the divergence when the blur ratio β does not improve anymore. A probabilistic way for optimizing both the selection and ordering of the set I of the frame pairs could arise by applying, e.g., probabilistic data association [28] to Delaunay triangle stars used in [4].

The data [29] used is a recording of a forest harvester operation [23]. Although the data allowed in developing some parts of the pipeline, crucial parts are missing. These are the aforementioned search for the fastest converging sequence of corrections, functional memory management of recent scanner views and a process converting individual views to a memoized global map [4], and countering possible systematical errors in relatively simplistic tree registration method presented in [23], which was used to generate the test data [29].

Since the *pcregistericp*() calls dominate (2800 initial matches edge-computed and 80 corrective matches versus 29 Go-ICP calls), the combined time stays tolerable and promising for a possible full implementation. The 2800 initial matches are to be edge-computed at the vehicle, and therefore this process is likely a subject of many optimizations concerning the sensors, application-specific integrated circuits (ASIC), and algorithmic developments [30].

The blur ratio β of Eq. 30 is very close to the dimensionality estimation by box counting [31]. The iteration starts with a box counting dimension estimate $d = 1.4$ and gets stagnated to $d \approx 1$ for a long time, while cluster arches of individual trees get shorter; see the top detail of Fig. 11. Then, dimensionality moves to the final $d = 0.3$. PC dimensionality would be a better iteration progress indicator in that it does not require specific parameters ϵ_1 and ϵ_2. Both indicators β and d apply in the 2D (projective maps) and 3D cases by a change of the power of ϵ in Eq. 30.

The search space reduction (230:1) shown in Table 3 has large but indirect effect to the computation, since the BnB process is hierarchical and eliminates large swath of search space rather soon. The reduction in computation time seems to be in the range of 3:1 ... 10:1, when uniformity assumptions are applied.

From the point of view of lightweight computation at the edge and cloud offloading in remote environments, the method we have proposed in this chapter presents some inherent benefits. First, by providing a self-corrective approach, there is potential to minimizing the drift in localization for autonomous mobile robots operating over large distances in places with a weak or missing Global Navigation Satellite System performance (GNSS-denied environments). For example, UAVs flying under tree canopy in forests for surveying applications that cannot rely on GNSS sensors are a potential application area. Second, by minimizing the size of the PC used for correcting the odometry process, we can provide cloud offloading or multi-robot collaboration even in environments where connectivity is poor and unreliable and latency does not allow for traditional computational offloading.

Therefore, large-scale maps can be built at the cloud or within multi-robot systems in remote environments.

Finally, it is worth mentioning that this method can be extended to multiple domains and application areas. From the perspective of the low computational complexity, this method can extend long-term autonomy in mobile robots by reducing the embedded hardware requirements. This is in turn related to lower energy consumption and applicability in smaller platforms. Moreover, if landmarks or anchors are well identified, this can also be leveraged within collaborative multi-robot systems, e.g., with micro-aerial vehicles being deployed from ground units in remote environments [32]. In forests environments in particular, our adaptive and lightweight self-corrective SLAM approach can be used for either canopy or tree stem registration, but other features that are distributed throughout the operational environments could be exploited as well.

The next two subsections are devoted to the discussion of alternative details of the Methods section.

4.1 Alternatives for Power Coefficients

The choice of power coefficient u_l seems to have great effect on the proposed iterative improvement scheme. Just as there are alternatives to the proposed interpolation scheme [33], there are alternatives for the formulation of u_l defined in Eq. 21:

1. Cumulative measures like the relative odometric path length $u_l = \sum_{k=j+1}^{l} \|q_k - q_{k-1}\| / \sum_{k=j+1}^{n} \|q_k - q_{k-1}\|$ of transformations $\tau_{k1}([omega_k], \theta_k, q_k)$ or accumulated match errors: $u_l = \sum_{k=j+1}^{l} e_k / \sum_{k=j+1}^{n} e_k$.
2. A more sophisticated SE(3) metric. One candidate is a linear combination of relative rotation and translation [34] $d(\tau_{lj}) = \sqrt{a\theta_{lj}^2 + b\|q_{lj}\|^2}$, where $\tau_{lj} = \tau_{lj}(\theta_{lj}, \vec{\omega}_{lj}, q_{lj})$ and $0 < a, b \in \mathbb{R}$ are free positive constants. This leads to
$$u_l = \frac{d(\tau_{lj})}{d(\tau_{ij})}. \tag{31}$$
If the scanner view cone is known, the above measure is very close to the mean squared distance between corresponding spots in the two view cones.
3. For paths with a lot of loops, one can find the nearest fit from the skew path $\{\tau_u \,|\, \tau_{j1}^{1-u} \tau_{i1}^{u}\}_{0 \le u \le 1}$ shown in Fig. 2:
$$u_l = \operatorname*{argmin}_u d(\tau_u^{-1} \tau_{lj}), \tag{32}$$
where $\tau_u^{-1} \tau_{lj}$ is a transformation from τ_{lj} to τ_u.

4.2 Frame Elimination

After the iterative improvement, some frames may show a large detrimental contribution to the final map quality. These frames can be removed. For this, one has to reshuffle the summation of the map error e_C to individual frames l, $1 \leq l \leq n$:

$$e_C^2 = \underset{i\in[1,n],p\in P_i{}'}{\text{mean}} \|p - c_{h_i}\|^2 = \sum_{i=1}^{n} w_i, \tag{33}$$

where an inclusion of $p \in P_i' \subset P_i$ occurs only if it contributes to some tree in the final map and w_i are the rearranged summand parts of the mean. The largest values can be removed. Finding a subset of frames to be removed is a combinatorially expensive operation, which should be done only if the application specifically requires it. Removing low-quality frames has similarities with the problem of selecting and ordering the corrective frame pairs, and both problems resemble feature selection over large feature space in general Machine Learning.

5 Conclusion

This article gives a complete presentation of mathematical details of rigid body interpolation and its application to iterative SLAM improvement. A main motivation was to provide a unified approach to the SLAM accuracy improvement. This resulted in an outline of the proposed iterative improvement algorithm. The second motivation was to test how much the very reliable Go-ICP algorithm can gain advantage from the small and sparse problems. It seems that Go-ICP is a feasible choice for tree map-related SLAM.

Our results suggest that the iterative SLAM improvement using rigid body interpolation proposed in this chapter has potential for many applications with sparse PCs, whether point clouds are key points, sets of beacons, or subsampled PCs. The near-uniform distribution makes the BnB search grid of Go-ICP coarser, and this and small PC size speed up Go-ICP, which is otherwise known to be a rather slow method. The sensor fusion with GNSS, inertial mass units, and other sensors has been left out to keep the presentation simple.

More research is needed especially about an optimal selection of the improvement matches before an effort to build a true pipeline from autonomous vehicle to a cloud environment can be done. The pipeline would cover the edge-computed tree registration and SLAM, transmission of sparse PCs to the cloud computing environment, and the iterative tree map improvement. This may take years but could be worth of an effort.

Acknowledgments The data was gathered in co-operation with Stora Enso Wood Suply Finland, Metsäteho Oy and Aalto University. The data collection was done under the EFFORTE, Efficient forestry for sustainable and cost-competitive bio-based industry (2016–2019) in WP3—Big data databases and applications.

References

1. Shan, T., & Englot, B. (2018). Lego-loam: Lightweight and ground-optimized lidar odometry and mapping on variable terrain. In *2018 IEEE/RSJ International Conference on Intelligent Robots and Systems (IROS)* (pp. 4758–4765). IEEE.
2. Qin, T., Li, P., & Shen, S. (2018). VINS-Mono: A robust and versatile monocular visual-inertial state estimator. *IEEE Transactions on Robotics, 34*(4), 1004–1020.
3. Queralta, J. P., Qingqing, L., Schiano, F., & Westerlund, T. (2020). VIO-UWB-based collaborative localization and dense scene reconstruction within heterogeneous multi-robot systems.
4. Li, Q., Nevalainen, P., Queralta, J. P., Heikkonen, J., & Westerlund, T. (2020). Localization in unstructured environments: Towards autonomous robots in forests with delaunay triangulation. *Remote Sensing, 12*(11), 1870.
5. Engel, J., Koltun, V., Cremers, D. (2018). Direct sparse odometry. *IEEE Transactions on Pattern Analysis and Machine Intelligence, 40*(3), 611–625.
6. Chang, L., Niu, X., & Liu, T. (2020). GNSS/IMU/ODO/LiDAR-SLAM integrated navigation system using IMU/ODO pre-integration. *Sensors, 20*(17), 4702.
7. Chetverikov, D., Svirko, D., Stepanov, D., & Krsek, P. (2002). The trimmed iterative closest point algorithm (Vol. 16, pp. 545–548).
8. Yang, J., Li, H., Campbell, D., & Jia, Y. (2015). Go-ICP: A globally optimal solution to 3D ICP point-set registration. *IEEE Transactions on Pattern Analysis and Machine Intelligence, 38*(1), 12.
9. The Mathworks, Inc. (2020). Natick, Massachusetts. *MATLAB version 9.8.0.1323502 (R2020a).*
10. Chen, Y., & Medioni, G. (1992). Object modelling by registration of multiple range images. *Image Vision Computing, 10*(3), 145–155.
11. Besl, P. J., McKay, N. D. (1992). A method for registration of 3-d shapes. *IEEE Transactions on Pattern Analysis and Machine Intelligence, 14*(2), 239–256.
12. Williams, B., Cummins, M., Neira, J., Newman, P., Reid, I., & Tardós, J. (2009). A comparison of loop closing techniques in monocular slam. *Robotics and Autonomous Systems, 57*, 1188–1197.
13. Wang, Z., Shen, Y., Cai, B., & Saleem, M. T. (2019). A brief review on loop closure detection with 3d point cloud. In *2019 IEEE International Conference on Real-time Computing and Robotics (RCAR)* (pp. 929–934).
14. Chisholm, R. A., Cui, J., Lum, S. K., & Chen, B. M. (2013). UAV lidar for below-canopy forest surveys. *Journal of Unmanned Vehicle Systems, 1*(01), 61–68.
15. Sankey, T., Donager, J., McVay, J., & Sankey, J. B. (2017). UAV lidar and hyperspectral fusion for forest monitoring in the southwestern USA. *Remote Sensing of Environment, 195*, 30–43.
16. Ortiz Arteaga, A., Scott, D., Boehm, J. (2019). Initial investigation of a low-cost automotive lidar system. In *ISPRS-International Archives of the Photogrammetry, Remote Sensing and Spatial Information Sciences* (Vol. 42, pp. 233–240). Copernicus GmbH.
17. Lin, J., & Zhang, F. (2020). Loam livox: A fast, robust, high-precision lidar odometry and mapping package for lidars of small FOV. In *2020 IEEE International Conference on Robotics and Automation (ICRA)* (pp. 3126–3131). IEEE.
18. Einhorn, E., & Gross, H. M. (2015). Generic NDT mapping in dynamic environments and its application for lifelong slam. *Robotics and Autonomous Systems, 69*, 28–39.

19. Lynch, K. M., & Park, F. C. (2017). *Modern robotics: Mechanics, planning, and control*. Cambridge University Press.
20. Moler, C., & Van Loan, C. (2003). Nineteen dubious ways to compute the exponential of a matrix, twenty-five years later. *Society for Industrial and Applied Mathematics, 45*, 3–49.
21. Dorst, L., Fontijne, D., & Mann, S. (2010). *Geometric algebra for computer science: An object-oriented approach to geometry*. San Francisco, CA: Morgan Kaufmann Publishers.
22. Dantam, N.T. (2018). Practical exponential coordinates using implicit dual quaternions. In *Workshop on the Algorithmic Foundations of Robotics* (pp. 639–655).
23. Nevalainen, P., Li, Q., Melkas, T., Riekki, K., Westerlund, T., & Heikkonen, J. (2020). Navigation and mapping in forest environment using sparse point clouds. *Remote Sensing, 12*(24), 4088.
24. Franceschetti, G., & Riccio, D. (2007). Chapter 2—surface classical models. In G. Franceschetti & D. Riccio (Eds.), *Scattering, Natural Surfaces, and Fractals* (pp. 21–59). Burlington: Academic Press.
25. Akkiraju, N., Edelsbrunner, H., Facelo, M., Mücke, E. P., Fu, P., & Varela, C. (1995). Alpha shapes: definition and software. In N. Amenta (Ed.), *Proc. Internat. Comput. Geom. Software Workshop*. Geometry Center Res. Rept..
26. Tang, J., Chen, Y., Kukko, A., Kaartinen, H., Jaakkola, A., Khoramshahi, E., Hakala, T., Hyyppä, J., Holopainen, M., & Hyyppä, H. (2015). SLAM aided stem mapping for forest inventory with small-footprint mobile LiDAR. *Forests, 6*(12), 4588–4606.
27. py-goicp implementation. Retrieved December 30, 2020 from https://pypi.org/project/py-goicp/
28. Bowman, S. L., Atanasov, N., Daniilidis, K., & Pappas, G. J. (2017). Probabilistic data association for semantic slam. In *2017 IEEE International Conference on Robotics and Automation (ICRA)* (pp. 1722–1729).
29. Nevalainen, P. (2020). Replication data for: Navigation and mapping in forest environment using sparse point clouds. https://doi.org/10.4225/13/511C71F8612C3
30. Liu, R., Yang, J., Chen, Y., & Zhao, W. (2019). eSLAM: An energy-efficient accelerator for real-time ORB-SLAM on FPGA platform. arXiv e-prints, arXiv:1906.05096.
31. Wu, J., Jin, X., Mi, S., & Tang, J. (2020). An effective method to compute the box-counting dimension based on the mathematical definition and intervals. *Results in Engineering, 6*, 100106.
32. Queralta, J. P., Taipalmaa, J., Pullinen, B. C., Sarker, V. K., Gia, T. N., Tenhunen, H., Gabbouj, M., Raitoharju, J., & Westerlund, T. (2020). Collaborative multi-robot search and rescue: Planning, coordination, perception and active vision. *IEEE Access, 8*, 191617–191643.
33. Žefran, M., & Kumar, V. (1998). Interpolation schemes for rigid body motions. *Computer-Aided Design, 30*(3), 179–189. Motion Design and Kinematics.
34. Park, F.C. (1995). Distance metrics on the rigid-body motions with applications to mechanism design. *Journal of Mechanical Design, 117(1)*, 48–54.

Future Possibilities and Challenges for UAV-Based Imaging Development in Smart Farming

Jere Kaivosoja

Abstract Technologies related to UAV (unmanned aerial vehicle) are developing rapidly. On the other hand, technologies related to farming are developing also, and several possibly revolutionizing technologies are about to become reality in farming. These technologies can set new goals and targets for the UAV imaging in smart farming. This work first reviews forthcoming technologies from measurement technologies, data management, execution technologies, and farming methods and then, as a top-down basis, formed possible imaging concepts for the future. The core future concepts were new imaging techniques with UAVs, data collection for digital twins and mapping for on-demand acting working UAVs and robotics. The presented technologies are at very early development stage.

Keywords Drone · Spraying · Hyperspectral · Multispectral · Mosaics · Concept · Precision farming

1 Introduction

Technologies related to UAV (unmanned aerial vehicle) imaging are developing rapidly, and there are already a wide variety of precision farming applications [1–3]. On top of the UAV development, other technologies related to farming are developing also, and several groundbreaking approaches might be met in sustainable intensification [4]. Those could also revolutionize the way UAV imaging is used in farming context in the future, thus setting new goals for the UAV development.

In precision farming, the goal is to spatially and timely optimize the farming inputs for maximizing the farming outcomes and minimizing the environmental load. This optimization is often difficult and a result of compromises. One of the starting points is to detect the current within-field variation. Remote sensing

J. Kaivosoja (✉)
Natural Resources Institute Finland (LUKE), Helsinki, Finland
e-mail: jere.kaivosoja@luke.fi

T. Lipping et al. (eds.), *New Developments and Environmental Applications of Drones*,
https://doi.org/10.1007/978-3-030-77860-6_6

can offer several approaches for that. The idea that combines all the relevant technologies in precision farming and in farm management is called smart farming. While precision farming goes into a single farming process one at the time, the smart farming covers the whole farming process [5, 6]. In short, smart farming in the context of arable farming is a combination of information management, precision farming knowledge, and farm machinery automation [7]. Smart farming is the key for sustainable farming in the future [8]. Currently digitalization and artificial intelligence (AI)-based farm management became the new norm, building basis for cyber-physical system (CPM) farm management approach [5]. These make it possible to exploit UAV imaging data in new ways.

The UAV technologies itself can also revolutionize the way we treat plants during the growing season. UAVs can measure the status of plants and detecting weeds or infested plants and estimating nutrient status [2, 3, 9–12] or other variations within the field without destroying the plants. The UAV flying patterns are mostly adapted from remote sensing based on aircrafts; there can be several ways to develop this. When imaging techniques and technologies are considered, their accuracy is a very important aspect. To access the measurement accuracy or more widely measurement quality, the true value, i.e., ground truth, is needed. The work [7] studies data quality aspects more thoroughly in the precision farming context. To estimate the ground truth for UAV measurements, external and possible independent measurements are needed. On top of UAV imaging, the application of spraying UAVs could change the way how we exploit measurement information since they can work on demand, up to daily basis. Examples of the concepts with spraying UAVs are widely provided in the paper [13]. However, the regulations of spraying UAVs in Europe are complex and incomplete and mostly restricting.

One more challenging aspect for the UAV imaging is the expected changes from monoculture, where a single crop or variety is cultivated in a field at a time. The need for multiple cropping [14] and intercropping methods has long been recognized. The UAV imaging is an interesting position where it could provide new tools for the intercropping management.

The objective of this study was to examine these different future trends of agriculture from the UAV imaging point of view to be able to underpin future possibilities and challenges. The main research question was that what kind of UAV imaging applications there will be in the future.

Similar to [15, 16], this study approaches its topic in a wide context. This work outlines future development strategies and presents possible application concepts. Due to rapid technology development around the topic, several suggested applications can be already developed to a certain technology readiness level (TRL) or even to a commercial stage in some cases. However, this paper examines the current UAV imaging applications in agriculture in contrast to development possibilities in future smart farming concept.

2 Material and Methods

The first step was a literature review on current UAV research, smart farming, digitalization, and agronomics to recognize future trends and development themes. The second step was to integrate findings from earlier empirical research case studies such as papers [10, 11, 17–19]. The following step was to form separate themes related to the future of UAV imaging in terms of a top-down approach. Figure 1 illustrates the general separate topics under development related to the UAV imaging in agriculture including main keywords of each.

Four separate main topics were recognized: (1) measurement technologies, (2) data management including communication technologies, (3) execution technologies, and (4) farming methods. These topics are next explained in more detail.

2.1 Measurement Technologies

The core of the UAV imaging is to produce high-quality measurement data for different purposes. According to paper [20], data quality in remote sensing data means spatial, radiometric, and spectral resolution, temporal resolution, precision in the cluster accuracy, positional accuracy, thematic precision, temporal validity, data completeness, spatial redundancy, readability, accessibility, and consistency.

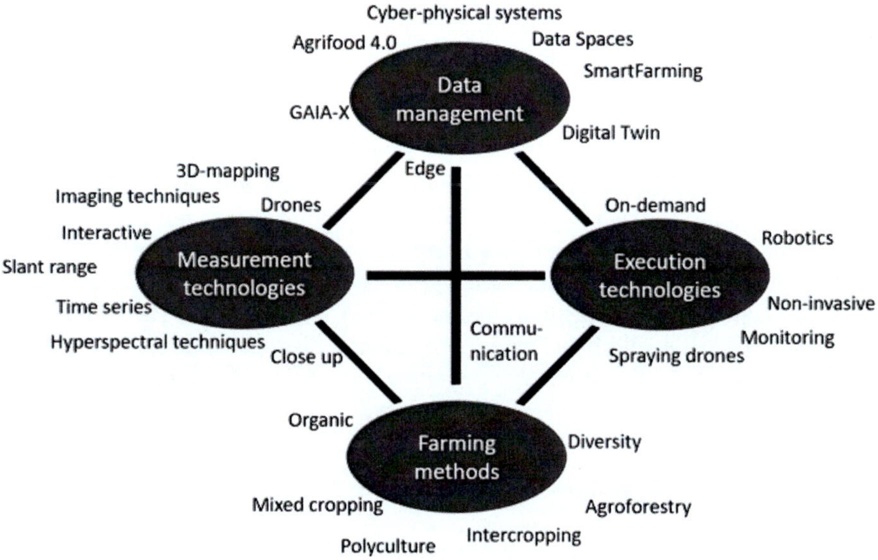

Fig. 1 General themes and related keywords around UAV imaging development scenarios in future agriculture

Fig. 2 Common pollen beetles on a rapeseed leaf being visible on a high-resolution image

These are also valid with UAV imaging data in agriculture when the purpose of data is clear. Instead of only the spatial redundancy, all image redundancies may be valid. Accessing feasibility of the developed imaging system requires fulfilling all the quality aspects, and that is often challenging. Typically agricultural cases construct orthophotos or use raw images. In study [21], the ratio was 70:30 to orthomosaics.

So far, the wider adaptation has concentrated on adapting simple flying patterns and measuring greenness differences and making direct precision farming application tasks based on it [9, 21]. New mapping methodologies should be developed [21, 22] to be able to see the targets from a better angle. For example, in the review [21], all the UAV imaging were done toward nadir. Selecting a suitable imaging system, namely, a camera, is also very important. Increasing spectral resolution might decrease spatial resolution in terms of applied tools that are available. Figure 2 presents common pollen beetles (*Meligethes aeneus*) on a leaf standing out easily on a high-resolution image. The optics in UAV imaging systems are often designed and optimized for orthophoto imaging from a distance of tens of meters [21].

The different sensor or technique integrations have provided promising results [11, 23], and that can be a way to overcome some of the challenges. The radiometric accuracy [17, 20] is still a challenge: to be able to distinguish target objects in a sunlight.

2.2 Data Management

Different machine learning [24] techniques are widely adopted to the UAV imaging. The main purpose of them is the successful classification of the imaged data. A typical challenge [21] is to gather enough reference data for teaching and validation. A fundamental deficiency is the lack of generalized solutions that work in different conditions. Computing solutions such as spectral unmixing [25, 26] can offer tools for improving the classification accuracy in specific cases when high spectral accuracy is applied. Current data analysis methods consider that imaging data is directly used or fused together with certain external data for the usage [27, 28]. Decision support system (DSS) (or AgriDSS) [29] is traditionally a software-based system which helps farmers to solve these issues. The farm management information systems (FMIS) are constantly developing [6, 15, 16, 30]. The adaptation of digital twin to the agriculture from industry, similar to adaptation of agriculture 4.0 [31], is interesting. Digital twin is a digital duplicate of the physical world, including the semantic object model, as well as the growth model that is characteristic for the specific crop and cultivar. In UAV imaging applications, digital twins mean that instead of deciding to put more fertilizers to less green areas, this makes it possible to measure plants and having a powerful system processing several data sources before making execution plans not directly related to imaging. Dataspace policies such as "toward a common European data space" [32] GAIA-X (Federated Data Infrastructure for Europe) and EFDI (Extended Farm Management Information Systems Data Interface) [33] are all supporting this digital twin-driven revolution by providing applicable data and processing strategies.

Communication technologies such as 5G, 6G, and beyond are key enablers together with IoT (Internet of Things) especially in the real-time interactions between external systems, satellites, and UAVs. Technologies such as edge computing may solve individual bottlenecks.

2.3 Execution Technologies

The technologies that carry out the farming actions are presented here as execution technologies. These technologies practically consider working machines at the field. The robotics development provides several options for UAV imaging or other in situ measurements. The UAV imaging is typically an additional task that needs to be done while field robots and tractors can measure while they work. This makes it questionable to develop UAV imaging for some cases, such as additional fertilization. However, one clear benefit from the remote sensing approach is that the overall needs can be estimated already before the actual work.

As presented earlier, the spraying UAVs offer new perspectives. Basically, they can work without destroying the cultivated plants or crops and without compacting the soil. This means that the spraying works can be made on demand, instead for an annual pesticide or herbicide spraying protecting against all possible threats. This means that the results of UAV imaging can be applied rapidly and more often. Also new types of tasks for UAVs such as pollinators [34] or biological controlling can offer tasks for imaging.

Fig. 3 Cropped image from a 3D model representing plots of different crops planned for intercropping near harvesting time. Small image represents relative greenness on the test site

2.4 Farming Methods

Currently there are several options for the typical monoculture, for example, mixed cropping, intercropping [35], and agroforestry [36]. Also methods such as organic farming and biodiversity friendly farming [37] can set new goals for UAV imaging. The crop row detections [38] applicable also for intercropping are fairly applied with UAV imaging [21]. Technologies such as beneath forest canopy navigating UAVs might become valid with agroforestry.

The monocultural farming is easier related to polycultural approaches especially from the harvesting point. Typical mixed cropping application is a mixed forage production [39]. Also the relation and interactions between different crops can vary a lot. Next Fig. 3. shows a 3D model based on DJI Phantom 4 RTK UAV imaging and DroneDeploy software. In the figure, several different crops potential for intercropping are grown in separate plots. Small image on Fig. 3. presents VARI (Visible Atmospherically Resistant Index) from the same plot setup showing relative greenness differences. The crops are reaching harvesting time, and the presented RGB colors and crop heights differ between plots. Selective harvesting [40, 41] is often needed in intercropping.

UAV imaging could, for example, reveal most suitable intercropping combinations in relation to harvesting, but the applicability of imaging in wide area intercropping is more challenging. The agroforestry can be seen as intercropping with trees; thus that approach would joint the current research themes of applications where same trees [42], vines [43], or shrubs are targets in different years.

3 Results

As the results of this study, different development possibilities and challenges are presented.

3.1 *Measurement Technologies*

The UAV operation can be developed to be more suitable for several agricultural aspects. The core is to provide better visibility of the target object:

- **Slant range:** Imaging with a tilted camera, often used with 3D modeling but with uneven canopy, the simple slanted angle may provide better target observation.
- **Close-up:** Close imaging may make small pests or insects visible. With limited coverage, sample points and interpolation can be applied.
- **Invasive:** Slightly moving the cover leaves can be required in some cases. Similar to robotics, the camera can be in the active part, but the data collection with UAVs needs to be done similar to close-up imaging.
- **Interactive:** This requires real-time analysis; the UAV can navigate to the interesting areas and go through them in detail.
- **Fleets:** Multiple drones can work interactively in a same case or in different cases. Fleets can decrease the field visit time.
- **Time series:** Autonomous UAVs can routinely inspect fields or targets, and the changes can be tracked similar to mining industries.

Possibilities for sensor technology development:

- **New technologies:** These are different spectral areas, fluorescence, and ground-penetrating radar. Each interesting phenomenon has different spectral response.
- **Active systems:** This will decrease errors in the pre-processing and could even provide absolute values like laboratory instruments, but it needs close-up imaging and controlled external lightning conditions.
- **New integrations:** Real-time integration to latest satellite imagery. Integration of different imaging solutions can ease processing.

These will have a great impact on all of the data quality elements.

3.2 *Data Management*

Impacts based on farming data management development:

- **Digital twin development:** Pre-processed data is recorded as such and it updates the digital twin. Imaging is not essential, but if done, it can be exploited. Measurements are not forced to estimate real values.

- **Simulation:** In addition of being part of digital twin, simulation will also take a role in reference data construction and provide synthetic data for measurement development.
- **Data availability:** This eases data analysis done by third-party, more specialized services
- **Internet of Things:** This provides real-time external data, such as information from traps. They can detect thresholds and launch UAV campaign or act interactively.
- **Communication technologies:** These enable real-time exploitation of external measurements.
- **Data processing:** Edge computing and real-time analysis offer more and more frequent results. Machine learning and new analyzing methods can offer robustness.

The affected main quality elements will be accessibility, readability, consistency, and data completeness. These data management technology developments will also have an impact on redundancies, thematic precision and cluster accuracy, and possibly temporal validity. Data management can also impact indirectly all of the quality elements.

3.3 Execution Technologies

Changes in current farming work executions may have the following effects:

- **Spraying UAVs**: non-destructive field works on demand; hotspot management
- **Other UAV works**: new applications for imaging: flowering, insects, etc.
- **Robotics support:** imaging to produce tasks for field robot in real time
- **Robotics overview:** scanning environment widely for robotics safety monitoring

Rather than impacting the quality elements, execution technologies will generate new imaging applications as such.

3.4 Farming Methods

The technologies related to farming methods are:

- **Intercropping technology:** plant optimization similar to precision farming, mixtures, and densities
- **Selective harvesting:** work execution timing at field scale or locally
- **Mixed field**: performance analysis during the growing season
- **Agroforestry:** Methods for accessing difficult locations

These methods will also rather deliver new applications than impact on quality.

4 Discussion and Conclusions

This paper presented several different ways how to develop UAV imaging in smart farming context. The time scale of these technologies varies a lot. While the working UAVs already exists, data management technologies are mostly concepts and general future frameworks and approaches. The related technologies were divided into four different categories listed here and followed by the most notable development branch: (1) measurement technologies, new imaging techniques and styles; (2) data management, digital twin and data collection for it; (3) execution technologies, working drones and on-demand actions; (4) farming methods, mapping the success of intercropping.

As this paper presented ideas at very early TRL levels, or even at the stage where the end technology is not recognized, the paper provided partly subjective view points as it is required when predicting future trends.

Two major challenges are recognized: the generalization of the imaging process and the economic payoff of different applications. As currently the UAV imaging is struggling with the affordability of precision farming actions, the billion-level impacts with pests and diseases are in hand.

One way for the next steps would be to go into details of individual cases by taking steps on TRL levels. Another possibility is to broaden the approach from the technology size to agronomy, agroecology, and economics and determine the most potential applications.

References

1. Kim, J., Kim, S., Ju, C., & Son, H. (2019). Unmanned aerial vehicles in agriculture: A review of perspective of platform, control, and applications. *IEEE Access, 7*, 105100–105115. https://doi.org/10.1109/ACCESS.2019.2932119.
2. Mogili, U. R., & Deepak, B. (2018). Review on application of drone Systems in Precision Agriculture. *Procedia Computer Science, 133*, 502–509.
3. Tsouros, D., Bibi, S., & Sarigiannidis, P. (2019). A review on UAV-based applications for precision agriculture. *Information, 10*(11), 349. https://doi.org/10.3390/info10110349.
4. Hunter, M., Smith, R., Schipanski, M., Atwood, L., & Mortensen, D. (2017). Agriculture in 2050: Recalibrating targets for sustainable intensification. *Bioscience, 67*(4), 385–390. https://doi.org/10.1093/biosci/bix010.
5. Wolfert, S., Ge, L., Verdouw, C., & Bogaardt, M. (2017). Big data in smart farming – A review. *Agricultural Systems, 153*, 69–80. https://doi.org/10.1016/j.agsy.2017.01.023.
6. Kruize, J., Wolfert, J., Scholten, H., Verdouw, C., Kassahun, A., & Beulens, A. (2016). A reference architecture for farm software ecosystems. *Computers and Electronics in Agriculture, 125*, 12–28. https://doi.org/10.1016/j.compag.2016.04.011.
7. Kaivosoja, J. (2019). *Role of spatial data uncertainty in execution of precision farming operations* (Aalto University publication series) (p. 66). Aalto University.
8. Walter, A., Finger, R., Huber, R., & Buchmann, N. (2017). Smart farming is key to developing sustainable agriculture. *Proceedings of the National Academy of Sciences of the United States of America, 114*(24), 6148–6150. https://doi.org/10.1073/pnas.1707462114.
9. Rasmussen, J., Nielsen, J., Streibig, J. C., Jensen, J. E., Pedersen, K. S., & Olsen, S. I. (2019). Pre-harvest weed mapping of Cirsium arvense in wheat and barley with off-the-shelf UAVs. *Precision Agriculture, 20*(5), 983–999. https://doi.org/10.1007/s11119-018-09625-7.

10. Oliveira, R. A., Nasi, R., Niemelainen, O., Nyholm, L., Alhonoja, K., Kaivosoja, J., et al. (2020). Machine learning estimators for the quantity and quality of grass swards used for silage production using drone-based imaging spectrometry and photogrammetry. *Remote Sensing of Environment, 246*, 111830. https://doi.org/10.1016/j.rse.2020.111830.
11. Viljanen, N., Honkavaara, E., Nasi, R., Hakala, T., Niemelainen, O., & Kaivosoja, J. (2018). A novel machine learning method for estimating biomass of grass swards using a photogrammetric canopy height model, images and vegetation indices captured by a drone. *Agriculture-Basel, 8*(5), 70. https://doi.org/10.3390/agriculture8050070.
12. Romero, M., Luo, Y., Su, B., & Fuentes, S. (2018). Vineyard water status estimation using multispectral imagery from an UAV platform and machine learning algorithms for irrigation scheduling management. *Computers and Electronics in Agriculture, 147*, 109–117. https://doi.org/10.1016/j.compag.2018.02.013.
13. He, X., Bonds, J., Herbst, A., & Langenakens, J. (2017). Resent development of unmanned aerial vehicle for plant protection in East Asia. *International Journal of Agricultural and Biological Engineering, 10*, 18–30.
14. Andrews, D., & Kassam, A. (1976). The importance of multiple cropping in increasing world food supplies. In R. I. Papendick, A. Sanchez, & G. B. Triplett (Eds.), *Multiple cropping* (ASA special publication 27) (pp. 1–10). Madison, WI: American Society of Agronomy.
15. Sorensen, C., Fountas, S., Nash, E., Pesonen, L., Bochtis, D., Pedersen, S., et al. (2010). Conceptual model of a future farm management information system. *Computers and Electronics in Agriculture, 72*(1), 37–47. https://doi.org/10.1016/j.compag.2010.02.003.
16. Fountas, S., Carli, G., Sorensen, C., Tsiropoulos, Z., Cavalaris, C., Vatsanidou, A., et al. (2015). Farm management information systems: Current situation and future perspectives. *Computers and Electronics in Agriculture, 115*, 40–50. https://doi.org/10.1016/j.compag.2015.05.011.
17. Honkavaara, E., Saari, H., Kaivosoja, J., Polonen, I., Hakala, T., Litkey, P., et al. (2013). Processing and assessment of spectrometric, stereoscopic imagery collected using a lightweight UAV spectral camera for precision agriculture. *Remote Sensing, 5*(10), 5006–5039. https://doi.org/10.3390/rs5105006.
18. Pesonen, L. A., Teye, F. K. W., Ronkainen, A. K., Koistinen, M. O., Kaivosoja, J. J., Suomi, P. F., et al. (2014). Cropinfra – An internet-based service infrastructure to support crop production in future farms. *Biosystems Engineering, 120*, 92–101. https://doi.org/10.1016/j.biosystemseng.2013.09.005.
19. Kaivosoja, J., Jackenkroll, M., Linkolehto, R., Weis, M., & Gerhards, R. (2014). Automatic control of farming operations based on spatial web services. *Computers and Electronics in Agriculture, 100*, 110–115. https://doi.org/10.1016/j.compag.2013.11.003.
20. Batini, C., Blaschke, T., Lang, S., Albrecht, F., Abdulm utalib, H., Basri, A., et al. (2017). Data quality in remote sensing. *The International Archives of the Photogrammetry, Remote Sensing and Spatial Information Sciences, Wuhan, China: SPRS Geospatial Week, XLII-2/W7*, 18–22.
21. Kaivosoja, J., Hautsalo, J., Heikkinen, J., Hiltunen, L., Ruuttunen, P., Näsi, R., Niemeläinen, O., Lemsalu, M., Honkavaara, E., & Salonen, J. (2021). Reference measurements in developing UAV systems for detecting pests, weeds and diseases. *MDPI Remote Sensing., 13*(7), 1238. https://doi.org/10.3390/rs13071238.
22. van der Merwe, D., Burchfield, D., Witt, T., Price, K., & Sharda, A. (2020). Chapter one – Drones in agriculture. In *Advances in agronomy* (pp. 1–30).
23. Nasi, R., Viljanen, N., Kaivosoja, J., Alhonoja, K., Hakala, T., Markelin, L., et al. (2018). Estimating biomass and nitrogen amount of barley and grass using UAV and aircraft based spectral and photogrammetric 3D features. *Remote Sensing, 10*(7), 1082. https://doi.org/10.3390/rs10071082.
24. Liakos, K., Busato, P., Moshou, D., Pearson, S., & Bochtis, D. (2018). Machine learning in agriculture: A review. *Sensors, 18*(8), 2674. https://doi.org/10.3390/s18082674.
25. Quintano, C., Fernandez-Manso, A., Shimabukuro, Y., & Pereira, G. (2012). Spectral unmixing. *International Journal of Remote Sensing, 33*(17), 5307–5340. https://doi.org/10.1080/01431161.2012.661095.

26. Yan, Y., Hua, W., Liu, X., Cui, Z., & Diao, D. (2019). Spatial-spectral preprocessing for spectral unmixing. *International Journal of Remote Sensing, 40*(4), 1357–1373. https://doi.org/10.1080/01431161.2018.1524590.
27. Kaivosoja, J., Pesonen, L., Kleemola, J., Pölönen, I., Salo, H., Honkavaara, E., et al. (2013). A case study of a precision fertilizer application task generation for wheat based on classified hyperspectral data from UAV combined with farm history data. In *SPIE remote sensing for agriculture, ecosystems, and hydrology XV2013*. SPIE.
28. Vanegas, F., Bratanov, D., Powell, K., Weiss, J., & Gonzalez, F. (2018). A novel methodology for improving plant pest surveillance in vineyards and crops using UAV-based hyperspectral and spatial data. *Sensors, 18*(1), 260. https://doi.org/10.3390/s18010260.
29. Lundstrom, C., & Lindblom, J. (2018). Considering farmers' situated knowledge of using agricultural decision support systems (AgriDSS) to Foster farming practices: The case of CropSAT. *Agricultural Systems, 159*, 9–20. https://doi.org/10.1016/j.agsy.2017.10.004.
30. Fountas, S., Sorensen, C., Tsiropoulos, Z., Cavalaris, C., Liakos, V., & Gemtos, T. (2015). Farm machinery management information system. *Computers and Electronics in Agriculture, 110*, 131–138. https://doi.org/10.1016/j.compag.2014.11.011.
31. Mirkouei, A. (2020). A cyber-physical analyzer system for precision agriculture. *Journal of Environmental Science: Current Research, 3*, 016.
32. Networks TD-GfC. (2020). Communication "towards a common European data space". In *Shaping Europe's digital future*. European Commission.
33. Olliver, A. (2017, February 25). *Powering precision farming with ISOBUS*. AXEMA-EurAgEng Conference, Villepinte, France.
34. Chechetka, S., Yu, Y., Tange, M., & Miyako, E. (2017). Materially engineered artificial pollinators. *Chempr, 2*(2), 224–239. https://doi.org/10.1016/j.chempr.2017.01.008.
35. Zhang, F., & Li, L. (2003). Using competitive and facilitative interactions in intercropping systems enhances crop productivity and nutrient-use efficiency. *Plant and Soil, 248*(1–2), 305–312. https://doi.org/10.1023/A:1022352229863.
36. del-Campo-Sanchez, A., Ballesteros, R., Hernandez-Lopez, D., Ortega, J. F., Moreno, M. A., & Agroforestry Cartography P. (2019). Quantifying the effect of Jacobiasca lybica pest on vineyards with UAVs by combining geometric and computer vision techniques. *PLoS One, 14*(4), e0215521. https://doi.org/10.1371/journal.pone.0215521.
37. Libran-Embid, F., Klaus, F., Tscharntke, T., & Grass, I. (2020). Unmanned aerial vehicles for biodiversity-friendly agricultural landscapes-A systematic review. *Science of the Total Environment, 732*. https://doi.org/10.1016/j.scitotenv.2020.139204.
38. Slaughter, D., Giles, D., & Downey, D. (2008). Autonomous robotic weed control systems: A review. *Computers and Electronics in Agriculture, 61*(1), 63–78. https://doi.org/10.1016/j.compag.2007.05.008.
39. Samarappuli, D., & Berti, M. (2018). Intercropping forage sorghum with maize is a promising alternative to maize silage for biogas production. *Journal of Cleaner Production, 194*, 515–524. https://doi.org/10.1016/j.jclepro.2018.05.083.
40. Leu, A., Razavi, M., Langstadtler, L., Ristic-Durrant, D., Raffel, H., Schenck, C., et al. (2017). Robotic green asparagus selective harvesting. *IEEE/ASME Transactions on Mechatronics, 22*(6), 2401–2410. https://doi.org/10.1109/TMECH.2017.2735861.
41. Lithourgidis, A., Dordas, C., Damalas, C., & Vlachostergios, D. (2011). Annual intercrops: An alternative pathway for sustainable agriculture. *Australian Journal of Crop Science, 5*(4), 396–410.
42. Imangholiloo, M., Saarinen, N., Markelin, L., Rosnell, T., Nasi, R., Hakala, T., et al. (2019). Characterizing seedling stands using leaf-off and leaf-on photogrammetric point clouds and hyperspectral imagery acquired from unmanned aerial vehicle. *Forests, 10*(5), 415. https://doi.org/10.3390/f10050415.
43. Kerkech, M., Hafiane, A., & Canals, R. (2020). Vine disease detection in UAV multispectral images using optimized image registration and deep learning segmentation approach. *Computers and Electronics in Agriculture, 174*, 105446. https://doi.org/10.1016/j.compag.2020.105446.

Role of Drones in Characterizing Soil Water Content in Open Field Cultivation

Antti Halla, Nathaniel Narra, and Tarmo Lipping

Abstract Soil water content is a central topic in open field cultivation. In Finland's boreal region with four thermal seasons, it has many roles which alter throughout the year. Climate change is changing the weather patterns, affecting all water-related processes and challenging the current farming practices. Better understanding of soils and their characteristics regarding response to water processes is called for, and data collection has a key role in this. Precision agriculture has been driving data intensification in farming. Unmanned aerial vehicles, or drones, have many applications and overall wide interest as an emerging technology in agriculture. Yet they lack an established role in day-to-day farming practices. Regarding data collection in open field cultivation, drones can be compared – or combined – with satellites, rovers, stationary devices, as well as plain old on-site observations by the farmer. In this study we give an overview of recent published literature, looking at data collection from the perspective of soil water information. We assess the opportunities and challenges of using drones in characterizing soil water content, mainly using soil and plant properties as proxies for it. Drones are useful in on-demand, nonintrusive, high-resolution spatial mapping of field properties. Soil moisture monitoring however requires frequent measurements, limiting the applicability of current drones.

Keywords Soil water content · Plant available water · Leaf reflectance · Soil reflectance

1 Introduction

Open field cultivation relies on a complex interdependent system of plants, soil, and weather. Water, essential for plant growth and the upkeep of soil biota, is a critical component in this system. Its movement between and within each part of the

A. Halla (✉) · N. Narra · T. Lipping
Tampere University, Pori, Finland
e-mail: antti.halla@tuni.fi; nathaniel.narra@tuni.fi; tarmo.lipping@tuni.fi

T. Lipping et al. (eds.), *New Developments and Environmental Applications of Drones*,
https://doi.org/10.1007/978-3-030-77860-6_7

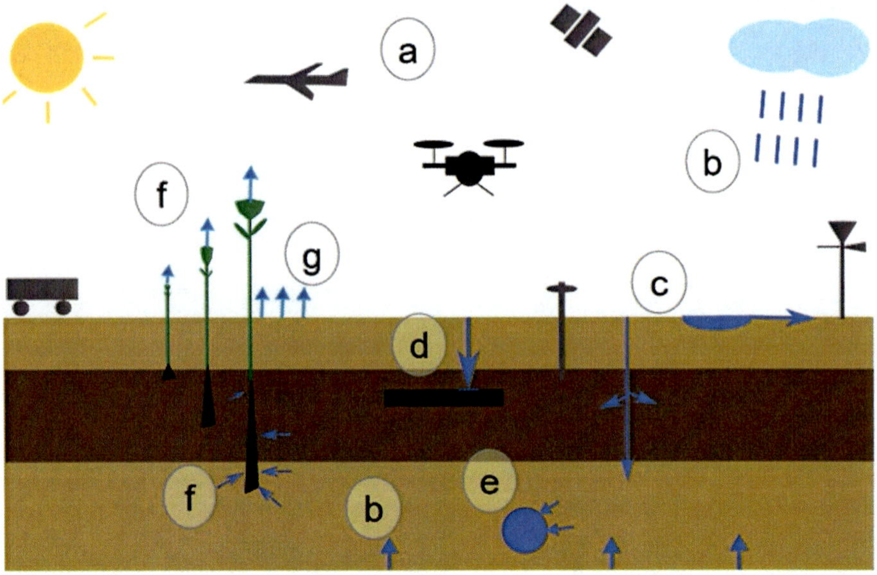

Fig. 1 Weather, soil, and plants form a dynamic system. (**a**) A combination of sensor platforms is needed for comprehensive measurements. (**b**) Soil water can be naturally recharged from above by precipitation and gravity and from below by ground water and capillary rise. (**c**) Rainwater needs to infiltrate the surface or it accumulates in puddles and flows overland. (**d**) Percolation to deeper layers may be prevented by a compacted hardpan. (**e**) Excess water is drained through underground drainage pipe. (**f**) Plants take in water through their roots and transpire water vapor through their leaves. (**g**) Part of the water is evaporated back to the atmosphere

system (Fig. 1) is guided by equally complex hydrological factors. Weather largely determines how much water enters a non-irrigated crop field and how much exits through evaporation loss. Water in the plants is essential for photosynthesis and other biological functions as well as for transport of soluble nutrients. Roots of most plants also need oxygen; therefore excess water at root level can damage the crops by creating anaerobic conditions.

Besides being crucial to plant growth plants, water can influence working conditions, soil biota, and soil morphology, all of which can in turn lead to lasting qualitative changes in the soil. Heavy machinery in wet, soft soil can lead to compaction in the soil [2], while freeze-thaw cycles can alleviate it [26].

Soil water can be naturally recharged from above by precipitation under gravitational force and from below by capillary forces on ground water. How the soil responds to incoming water is defined by soil hydrological properties which are largely a function of mineral texture, aggregate structure, and soil organic content. These properties show significant spatial variation at scales small enough to show heterogenous distributions between fields and even within individual fields, both horizontally and vertically. A combination of multiple sensor platforms and measurements help in forming a complete picture of the water dynamics within a field: satellites, aircrafts, ground vehicles, weather stations, ground probes, and drones.

1.1 The Need for Water Data

Knowing the current and forecasted state of soil water content and its availability to plants has significant implications to optimizing resources. Indeed, an ecologically sound management of the farm requires this information. Understanding the dynamics of this soil-water-plant system requires data that can characterize it and help explain the underlying processes. As this understanding develops, more needs for additional data collection can be identified. Improvements in technology enable us to measure properties of this system that has previously been out of reach. They can also increase the quantity and quality of the collected data in general as well as the efficiency of the data collection process itself.

Climate change is expected to disrupt water-related phenomena. The outlook in western Finland is more precipitation from autumn to spring, less snow coverage, longer dry periods during growing season although more intense rainfall events [39]. Predicting water content in different weather scenarios can help in assessing the risk to crop production. The current crop growth models in use that depend on soil moisture content are typically calibrated for regional water patterns of recent history. If these patterns are to change, then it becomes essential to recalibrate management practices. In addition, increasing awareness of and drive to end inefficient practices that lead to nutrient leaching, or water wastage, has made soil moisture monitoring an important aspect in open field farming.

1.2 Measuring Soil Water Content

Soil hydrology and measurement of true soil water content have been presented comprehensively by Novák et al. [37]. While providing accuracy these methods are labor, time, and resource intensive, which limit their practical application in everyday farming context. Notably, the *gravimetric method* requires extracting a soil sample and drying it in an oven. They do however serve an important role in ascertaining the true value with some confidence, in instrument calibration and model development. Data on features whose correlation with soil-water is established can also be used, albeit with lower confidence. If a correlation can be established, there is value in low fidelity characterization of soil-water with high acquisition ease and resolution. Remote sensing methods together with the necessary data post-processing steps can be computationally more intensive and also laborious to set up but are more conducive for automation.

Water content varies with space and time. Spatial variation is, for example, due to differences in soil properties, topology, precipitation level, and relative location respective to underground drainage, which makes reliable extrapolation from point measurements difficult. The rate of change can be high especially during growing season.

In this study, instead of direct measurement, we look at the most common methods for estimating the soil water content through indirect proxy measurements that correlate with the actual soil moisture. For the purposes of this study, we divide these proxy variables into *soil proxies* and *plant proxies*, which measure soil water content through soil or plant properties, respectively. Soil proxies measure properties of the bare ground, such as its dielectric permittivity or spectral reflectance. Plant proxies measure, for example, the plant canopy's spectral reflectance or temperature. The list of variables here is not meant to be exhaustive. Any variable that correlates with soil water content could be included. Well-known examples are given while the focus of the study is in the methodology.

A conceptual model is presented in Fig. 2 describing how, for the purposes of this study, the water-related phenomena and variables are assumed to be related to the measurements of proxy indicators. As an example, the Normalized Difference Water Index NDWI (sometimes called moisture index NDMI) uses two infrared spectral bands of an image sensor to measure liquid water content of vegetation canopies. Similarly, the commonly used vegetation index NDVI is calculated from red and near-infrared channels, measuring chlorophyll-related leaf reflectance in the plants [20]. Plant available water is a necessary, although not sufficient, condition for the existence of chlorophyll in leaves. Thus, both NDWI and NDVI are plant-related proxies for plant available water. However, as virtually all proxies, they are influenced by many other factors. Two main factors, namely, soil characteristics and plant type, are shown in the figure as *parameters*, which may need to be calibrated for when interpreting proxy measurements.

The model in Fig. 2 is static, depicting the state of the system at a given moment. Each arrow represents an assumed causal influence in the direction of the arrow. This framework is used as a guiding structure for organizing the literature overview in this study.

1.3 Drones in Agriculture

The use of drones, or more formally – unmanned aerial vehicles (UAVs) – in agriculture is increasing rapidly. UAVs are used in crop production, forestry, and disaster risk reduction. Applications in crop production range from crop health monitoring and irrigation planning to weed detection and insurance [17].

Broadly speaking, the role of drones can be seen either as passive or active, depending on how much they interact with their environment. Active drones can spray pesticides or take physical samples. This study is focused on passive data collection. The actual end application of the data can be crop stress monitoring, crop detection, seasonal planning, loss estimation, etc. In these roles, drones can be seen as an alternative or a complementary tool for acquiring the needed information.

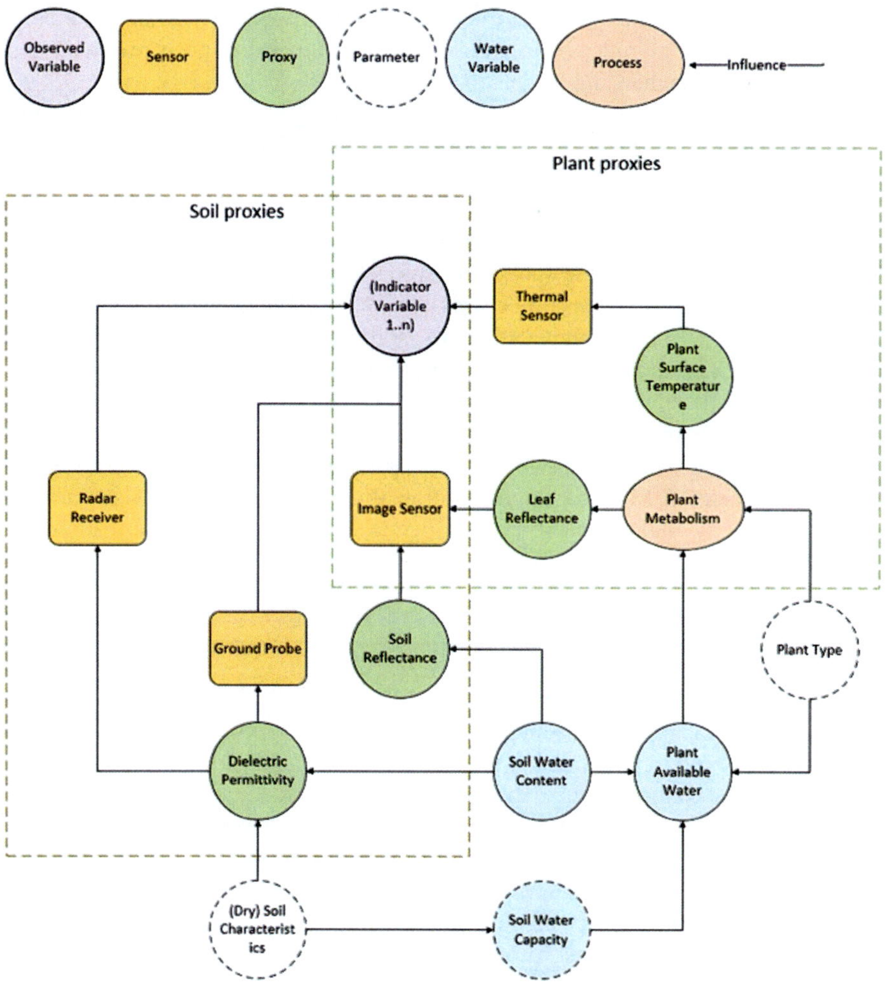

Fig. 2 Soil and plant properties as proxies for water related information in the soil. Measurements are typically several steps apart from the actual phenomena of interest. There are several paths that can be taken, depending on the situation, available technology, and resources. Outputs from multiple sensors can be combined for more reliable indicators

In addition, drones have been used in ancillary roles, such as in collecting data from wireless sensor networks [42, 47].

Drones, ground vehicles, airplanes, and satellites can all be equipped with same or similar measurement technology. Therefore, much of the literature regarding soil water measurement is presented under the topic of the given technology. Data acquisition by drones typically has specific advantages and constraints in terms of spatial and temporal resolution of the data as well as practical considerations of operating a drone. The wide and growing range of different types and sizes of drones from large military drones to "smart dust"; their properties in terms of endurance, range, weight, and altitude; as well as their applications has been reviewed and classified by Hassanalian and Abdelkefi [24].

The ability to do on-demand high-resolution mapping of fields without disturbing either soil or plants is a main advantage of a drone in open-field farming. Constraints on a drone as a sensor carrier platform are due to factors such as weight of the instruments, required proximity to ground, and flight time. Airborne vehicles measuring soil and canopy surface only through reflectance and spectrometry have limitations. Some of the sub-surface properties can be estimated from these images, if the data generating processes are known. Using manual sampling to acquire reference data on parameters such as soil texture, soil structure, and organic content, for example, would allow for more accurate estimates compared to visual examination of the obtained images alone.

Drones themselves are becoming commodity items. The value of drones in data collection however would come from their capabilities and limitations as a sensor-carrying platform. Utility for a farmer additionally requires a mature data processing workflow that can routinely turn the raw collected data into relevant information that improves their field management decisions.

1.4 Objectives

This is a preliminary study to map out the different aspects related to measuring water content in a crop field and to provide an overview of recent literature on the topic. We approach the crop field as system from two different perspectives – water state and soil properties. We look at methods to measure the water state on the field at a given moment, as a snapshot of the dynamic system. That enables water stress detection in monitored conditions. We also look at measuring system properties, those relatively stable soil hydrologic properties that could aid in characterization of soil's response to water events. Knowledge of this response would allow us to better estimate the soil moisture conditions at a given time, even with scarce data. It would also enable forecasting the availability of water to plants and simulate the behavior of the field in different scenarios such as prolonged drought or excess rainfall. We look at how drones have been used in these tasks and what is the future outlook.

2 Measurement Targets

The target is to estimate the actual water content in the soil and the plant available water specifically. As discussed above and shown in Fig. 2, we focus on measuring the variables of interests indirectly through proxies in soil and plants. There are also other variables, such as weather data, that covary with soil water content and could therefore be used to improve estimations of it.

2.1 Soil Proxies

Spectral properties of soil correlate with soil moisture and can be used to monitor moisture conditions in bare soil [16], such as drying of the soil in spring. Remote sensing soil surface reflectance has limited utility as a proxy, as it provides information mostly about the soil surface, which can differ considerably from moisture below. Soil surface characteristics in itself have a large role in determining the amount of infiltration and runoff, especially with crusting [11]. Furthermore, even the soil surface is hidden below the crop canopy during the growing season.

Soil relative dielectric permittivity is a typical soil proxy and can be measured using ground probes [37] as well as radar [9, 30].

When there is plant canopy, the soil is typically not directly observable using reflectometry, but a soil probe or radar is needed.

2.2 Plant Proxies

Plant physiological properties can be used to detect water stress, which in turn is an indicator of plant available water content at root level. Gago et al. [19] reviewed literature on using UAVs to measure water stress by using leaf reflectance and temperature and called for studies measuring leaf chlorophyll fluorescence as a more direct indicator of photosynthesis.

Spectral imaging, both within and beyond the visible spectrum, is the main method for measuring plant proxies. Information from different spectral bands is combined into specialized indices that target specific phenomena of interest, for example, NDVI for crop vegetation vigor, NDWI for water content of crop canopy [20], and Leaf Area Index (LAI) for estimating evapotranspiration among other properties [48]. Candiago et al. evaluated vegetation indices for precision farming applications from multispectral UAV images in [7].

Hassan-Esfahani et al. [25] used plant proxy together with gravimetric reference samples to produce a machine learning model for surface soil moisture estimation. In addition to multispectral and thermal images, their best performing input combinations required knowledge of field capacity, derived from soil texture samples.

2.3 Soil Characteristics

Soil characteristics are those relatively stable properties of the soil that affect the interaction between soil and water, i.e., how rainfall infiltrates the surface, percolates through soil layers, how much of it is retained in the soil, and how much available for plants to use [37]. Mineral texture and aggregate structure of the soil as well as its organic matter content are the main components. These are relatively stable properties, which can be considered constant at least within a single season. There can however be considerable spatial variation within a field, both on the surface and sub-surface. The presence, type, and condition of underground drainage systems and other installed infrastructure that affect the water dynamics of the field are similarly stable factors in the system.

Soil characteristics don't generally vary with water content, but knowledge of them can aid in the estimation of water content and its behavior over time, as they affect the sensor measurements and may need to be calibrated for. They can also be used as inputs to *pedotransfer functions* (PTF) to estimate soil hydrological properties, such as field capacity and wilting point, based on their statistical relationships in large soil data sets [43]. The traditional way to characterize soils is by collecting soil samples and analyzing them in a laboratory. Technologies that would measure soil properties in situ or reducing the number of needed soil samples would therefore also help in estimating soil water content.

3 Measurement Technology

Technology for measuring proxies for soil water content in soil and plants is divided here into image sensors, radar, and ground probes. As same or similar sensors can be mounted on different platforms, sensor-specific reviews irrespective of platform are also included here for broad perspective and summarized in Table 1. Selected case studies that exhibit utility of drones as a platform are summarized in Table 2. An overview on the literature is given briefly below.

3.1 Ground Probes

Ground measurements can be used to collect continuous time series data such as soil moisture data with sub-surface sensors. Calibration can be an issue, as there is no generic method that works across all manufacturers. Many manufacturers don't give access to either to the raw data or the internally applied conversion functions [27].

One challenge is that ground probes measure point data. This point data needs to be interpolated over potentially highly variable soil properties. In addition to point sensors, mobile sensors for apparent soil electrical conductivity have been used for soil spatial variability mapping [12].

The use of drones here is limited. Using drone mounted ground probes would require a mechanism for inserting the probe into the soil without damaging it. The soil around the inserted probe also needs time to settle for accurate reading. Otherwise, drones could be used to collect data from sensors with wireless connectivity or in combination with a ground vehicle.

Table 1 Generic sensor related reviews

References	Title	Sensor	Year
Gago et al. (2015)	UAVs challenge to assess water stress for sustainable agriculture	RGB, multispectral, hyperspectral, thermal	2015
Barbedo (2019)	A review on the use of unmanned aerial vehicles and imaging sensors for monitoring and assessing plant stresses	RGB, multispectral, hyperspectral, thermal	2019
Adão et al. (2017)	Hyperspectral imaging: a review on UAV-based sensors, data processing and applications for agriculture and forestry	Hyperspectral	2017
Lu et al. (2020)	Recent advances of hyperspectral imaging technology and applications in agriculture	Hyperspectral	2020
Messina and Modica (2020)	Applications of UAV thermal imagery in precision agriculture: state of the art and future research outlook	Thermal	2020
(Brocca et al., 2017)	A Review of the Applications of ASCAT Soil Moisture Products	Radar (ASCAT)	2017
(Edokossi et al., 2020)	GNSS-reflectometry and remote sensing of soil moisture: a review of measurement techniques, methods, and applications	Radar (GNSS-R)	2020
Liu et al. (2016)	Ground penetrating radar for underground sensing in agriculture: a review	GPR	2016
Klotzsche et al. (2018)	Measuring soil water content with ground penetrating radar: a decade of progress	GPR	2018
Zajícová and Chuman (2019)	Application of ground penetrating radar methods in soil studies: a review	GPR	2019
Corwin and Scudiero (2020)	Field-scale apparent soil electrical conductivity	Apparent Electrical Conductivity sensor	2020
Babaeian et al. (2019)	Ground, proximal, and satellite remote sensing of soil moisture	Ground, proximal, satellite	2019
Hardie (2020)	Review of novel and emerging proximal soil moisture sensors for use in agriculture. sensors	Proximal sensors	2020
Jackisch et al. (2020)	Soil moisture and matric potential – an open field comparison of sensor systems	Ground probe	2020

3.2 Radar

Radar is an active remote sensing method, where the device sends an electromagnetic pulse and records the reflected or scattered return wave. Depending on the wavelength, radar signal can penetrate the plant canopy or soil. Radar signal frequencies range from high frequency radio waves (MHz) to around 100 GHz

Table 2 Selected case studies, estimating soil water content using UAV mountable technology through proxies in soil and plants or characterizing stable, water related soil properties

References	Title	Sensors	Metrics	Target	Year
Hassan-Esfahani et al. (2015)	Assessment of surface soil moisture using high-resolution multi-spectral imagery and artificial neural networks	RGB, NIR, thermal	Surface soil moisture	Plant proxy	2015
Matese et al. (2015)	Intercomparison of UAV, aircraft and satellite remote sensing platforms for precision viticulture	Multispectral	NDVI intra-vineyard vegetation variability	Plant proxy	2015
Christiansen et al. (2017)	Designing and testing a UAV mapping system for agricultural field surveying	Lidar	Crop height and volume	Plant proxy	2017
Ge et al. (2019)	Combining UAV-based hyperspectral imagery and machine learning algorithms for soil moisture content monitoring	Hyperspectral	Soil moisture content monitoring	Plant proxy	2019
Wu et al. (2019)	A new drone-borne GPR for soil moisture mapping	GPR	Soil moisture mapping	Soil proxy	2019
Chantasen et al. (2020)	Mapping the physical and dielectric properties of layered soil using short-time matrix pencil method-based ground-penetrating radar	GPR	Dielectric properties, dielectric constant, bulk density	Soil	2020

microwaves. Soil moisture can be derived from the dielectric constant extracted from the return signal [31].

Advanced Scatterometer (ASCAT) on board of satellites has been used in measuring surface soil moisture but is in its own not enough for precision agriculture applications [6] due to coarse (1 km+) resolution.

Global Navigation Satellite System Reflectometry (GNSS-R) is a method that combines active satellite signals with an on- or near-ground passive receiver and has been used to measure surface soil moisture [14]. A prototype for UAV mounted GNSS-R for retrieving soil moisture was presented by Jia et al. [28].

Ground-penetrating radar (GPR) is a term used for radars capable of subsurface measurements, with signals in 10–2000 MHz range. They have been used in civil engineering [44] and archeology [8] for applications such as object detection and assessing structural health. Resolution and maximum measurement depth depend on the signal frequency: higher frequency detects finer details but attenuates faster. In agriculture the depth of interest is approximately from the surface down to the depth of 2 meters. With a proper frequency, it is possible to characterize the soil in agriculture-relevant depth [32].

GPR has been used in characterizing soil characteristics and dielectric properties [9]. It can also give information about the soil water content, which affects the dielectric permittivity of the soil, which in turn affects the radar signal [30]. Zajícová and Chuman [46] reviewed applications of GPR in soil studies, concluding that it can assist in estimating soil moisture and in detecting soil horizons, especially in sandy soils with low cation exchange capacity (CEC). They point out that while clay soils have been found to be unfavorable for GPR surveys, more due to CEC than grain size, with high attenuation of the radar signal, even a penetration depth of 0.5 m would be sufficient for many applications.

GPR has been mounted on drones for landmine detection by Fernández et al. in [21], using synthetic aperture radar (SAR) algorithm. Wu et al. [45] described using a 1.5 kg radar system operating at 500–700 MHz range to map moisture in the top 10–20 cm of the soil.

3.3 Spectral Imagery

Cameras are typically categorized as visible light (RGB), multispectral, hyperspectral, and thermal cameras depending on the electromagnetic spectral range within which the sensors operate. RGB cameras operate within the visible spectrum (red, green, and blue spectral bands). Multispectral cameras include the visible range as well as selected bands in the near-infrared (NIR) short-wave infrared (SWIR) range. Hyperspectral cameras work in the same spectral range, but while both RGB and multispectral cameras capture distinct spectral bands, hyperspectral cameras capture a contiguous spectral range. A basic premise in spectral imagery is that certain plant phenological phenomena can be correlated to reflectance values in specific wavelengths. These wavelengths, or bands, can be further refined into specialized indices. For example, NDWI is a combination of two near-infrared channels which correlates with liquid water molecules in vegetation canopy [20].

Drone mounted hyperspectral cameras were reviewed by Adão et al. [1]. They saw potential for drones as platform as the hyperspectral devices are becoming smaller and lighter, although the amount of data collected by these devices can be huge and the required processing complex. Hyperspectral camera was used by Ewing et al. for soil gradation in laboratory conditions [15]. Hyperspectral devices have previously been out of reach for many farmers due to high price, but more affordable technologies and options are being developed and productized. One approach is combining regular digital cameras with passive diffraction grating filter and machine learning [41]. Even open-source do-it-yourself cameras have been built and tested, such as in [40].

Barbedo reviewed the use of drone imaging for monitoring plant stresses. He concluded that all approaches for water stress detection found had limitations for practical adoption. Combining data from multiple complementary sources was seen as the way forward, along with improved sensor technology, computer vision, and machine learning techniques [5].

Thermal cameras, operating mostly in the mid-infrared wavelengths (3–8 μm), have been used for assessing water stress with Crop Water Stress Index (CWSI). This and other promising applications of UAV thermal imagery such as subsurface drainage mapping are reviewed by Messina and Modica [36]. They pointed out that low-resolution compared to RGB images, low number of applications for thermal data, and required knowledge of thermography in the process are all limiting the adoption of thermal cameras. López and Giraldo proposed a method for planning an optimal irrigation route and rate, based on CWSI [33].

Drones have been used to provide variability maps to produce better extrapolations from point measurements. They have been used when the availability of satellite imagery has been limited or when the spatial resolution hasn't been high enough. The minimum pixel size for multispectral imagery of the Sentinel 2 satellite from European Space Agency is 10x10m [3], while drones can achieve sub-centimeter resolution.

3.4 Other Measurement Technologies

Light detection and ranging (LiDAR) has been used to create digital elevation models [38] and to map canopy volume and height for biomass [10]. This data can be used as proxies for crop vigor which in turn can be used with models for field water balance, water flows, and accumulation points. Fitzpatrick et al. proposed using thermoacoustic imaging that combines a microwave source with an ultrasound receiver to overcome some of the limitations of current technology, especially GPR [18]. Hardie reviewed a range of soil moisture sensors for use in agriculture, identifying their limitations and concluding that current technology for soil moisture measurement generally doesn't often meet the practical requirements in farming [23]. Babaeian et al. did an extensive review on ground, proximal, and satellite remote sensing of soil moisture, touching also other methods such as neutron scattering, nuclear magnetic resonance, and gamma ray sensors [4]. They found that many common methods only measure surface soil moisture and then proceeded by reviewing modeling approaches for root zone soil moisture estimation.

4 Discussion

Measuring plant available water content at the field scale with confidence requires combining data from multiple data sources. Making conclusive inferences about soil water content based solely on remote sensing data can be difficult, if not impossible. This uncertainty in the estimations could be reduced by combining the empirical models for spectral data with physically based models such as hydrological and crop growth models [4]. The cause of the detected features can be hard to identify because many causes can lead to similar data patterns. While ground probes provide point data with comparatively more proximal sensing of data, remote sensing technology can be used for gaining spatial distribution information, and drones can be mounted with several types of instruments to measure additional data on demand.

Opportunities for using drones in information collection are found when looking at the gaps left by other technologies and seeing how the strengths of drones such as spatial resolution, adaptability for different instruments, and non-intrusiveness can help. Satellites with 10 m + spatial resolution can provide approximate information at field level, but within-field variability may require higher resolution, which can be acquired using drone-mounted cameras. For example, estimation of LAI using Sentinel-2 data was studied and found to be unsatisfactory especially in precision agriculture when within-field variability is a concern [29].

Satellite based radars (C-band, L-band) can typically measure soil to a vertical depth of 2–7 cm of surface soil, but their horizontal surface resolution can be above 1 km [6]. On the other hand, near-surface ground-penetrating radars can characterize soils in deeper layers with much higher resolution, both vertically and horizontally [9, 30, 45, 46]. UAVs have the advantage of non-intrusiveness which allows it to be used during growing season to get information about subsurface conditions while also enables quick mapping of the whole field. High spatial resolution opens up the possibility of measuring chlorophyll fluorescence which could provide a more direct indicator of photosynthesis and allow for detection of water stress [19].

The conceptual model presented in the introduction is a static model with the main purpose of guiding the literature research. It could be further extended for use as a basis for a numerical model for estimating individual unknown variables, when other variables are known. Each arrow would then add uncertainty to the estimate as proxies move further away from the actual measurement target. For a useful dynamic model, weather and evapotranspiration would need to be included as well.

All the different methods of measuring soil moisture described here are based on indirect measurements of proxy variables and therefore unlikely to be satisfactory individually. They are subject to various degrees of uncertainty from multiple different sources, including measurement error and model error. When proceeding from estimation of current state to forecasting, this uncertainty will be further amplified. An important task is to quantify the different sources of uncertainty and their contribution to the overall prediction. This uncertainty can then be reduced by targeted data collection, including complementary measurements from different types of sensors [13].

5 Conclusions

The interaction of soil and water is a central topic in open field farming, and the changing weather patterns call for re-evaluating current field management practices. Measuring soil water content is a multi-faceted spatiotemporal problem with varying degrees of uncertainty. A tractable solution can be to measure soil properties and assess plant canopy condition. Soil characteristics define the more stable part of the system, while the highly variable water content can be estimated through plant- and soil-related proxies.

In this study, several key variables related to soil and plant available water content along with their relationships were identified and presented as a conceptual graph model. This graph was then used to guide a literature search to find applications of drones in measuring soil water content and characteristics in open field cultivation. This graph can be used as a tool to choose measurement methods and targets and diagnostics and in understanding the field water dynamics in general. It could further be used as a basis for developing a model to estimate the yield gap on a given field due to water issues. In the future this graph can be iteratively expanded upon as the need to include additional features and processes to the model arises.

Drones have characteristics that make them useful in measuring water content in open field cultivation, especially in on-demand, high-resolution spatial mapping. Visual range and multispectral cameras are commonplace in drones, but hyperspectral and thermal cameras along with radar technology can be mounted on drones as well, taking into account weight and cost limitations in practical applications.

Continuous monitoring and forecasting soil moisture require frequent measurements. Autonomous drones may be able to do this in the future, but at the moment this information needs to be provided by ground probes and other devices deployed at the field. In these cases, drones may still have a role in measuring the more stable properties and patterns of the field that can be used to improve the soil moisture estimates.

Even with collected data – with or without drones – the challenge of combining and analyzing data from different sources remains, before the results can be useful in practical decision-making for a farmer. The amount of data collected and the number of potentially computationally intensive, post-processing steps required can quickly become overwhelming, calling for proper infrastructure and high level of automation throughout the data processing chain.

References

1. Adão, T., et al. (2017). Hyperspectral imaging: A review on UAV-based sensors, data processing and applications for agriculture and forestry. *Remote Sensing, 9*(11), 1110. https://doi.org/10.3390/rs9111110
2. Alaoui, A., & Diserens, E. (2018). Mapping soil compaction – A review. *Current Opinion in Environmental Science & Health, 5*, 60–66. https://doi.org/10.1016/j.coesh.2018.05.003
3. Ambrosone, M., et al. (2020). Retrieving soil moisture in rainfed and irrigated fields using Sentinel-2 observations and a modified OPTRAM approach. *International Journal of Applied Earth Observation and Geoinformation, 89*, 102113. https://doi.org/10.1016/j.jag.2020.102113
4. Babaeian, E., et al. (2019). Ground, proximal, and satellite remote sensing of soil moisture. *Reviews of Geophysics, 57*(2), 530–616. https://doi.org/10.1029/2018RG000618
5. Barbedo, J. G. A. (2019). A review on the use of unmanned aerial vehicles and imaging sensors for monitoring and assessing plant stresses. *Drones, 3*(2), 40. https://doi.org/10.3390/drones3020040
6. Brocca, L., et al. (2017). A review of the applications of ASCAT soil moisture products. *IEEE Journal of Selected Topics in Applied Earth Observations and Remote Sensing, 10*(5), 2285–2306. https://doi.org/10.1109/JSTARS.2017.2651140
7. Candiago, S., et al. (2015). Evaluating multispectral images and vegetation indices for precision farming applications from UAV images. *Remote Sensing, 7*(4), 4026–4047. https://doi.org/10.3390/rs70404026
8. Catapano, I., et al. (2019). Applying ground-penetrating radar and microwave tomography data processing in cultural heritage: State of the art and future trends. *IEEE Signal Processing Magazine, 36*(4), 53–61. https://doi.org/10.1109/MSP.2019.2895121
9. Chantasen, N., et al. (2020). Mapping the physical and dielectric properties of layered soil using short-time matrix pencil method-based ground-penetrating radar. *IEEE Access, 8*, 105610–105621. https://doi.org/10.1109/ACCESS.2020.2999894
10. Christiansen, M., et al. (2017). Designing and testing a UAV mapping system for agricultural field surveying. *Sensors, 17*(12), 2703. https://doi.org/10.3390/s17122703
11. Corbane, C., et al. (2012). Multitemporal analysis of hydrological soil surface characteristics using aerial photos: A case study on a Mediterranean vineyard. *International Journal of Applied Earth Observation and Geoinformation, 18*, 356–367. https://doi.org/10.1016/j.jag.2012.03.009
12. Corwin, D. L., & Scudiero, E. (2020). Field-scale apparent soil electrical conductivity. *Soil Science Society of America Journal, 84*(5), 1405–1441. https://doi.org/10.1002/saj2.20153
13. Dietze, M. C. (2017). Prediction in ecology: A first-principles framework. *Ecological Applications, 27*(7), 2048–2060. https://doi.org/10.1002/eap.1589
14. Edokossi, K., et al. (2020). GNSS-Reflectometry and remote sensing of soil moisture: A review of measurement techniques, methods, and applications. *Remote Sensing, 12*(4), 614. https://doi.org/10.3390/rs12040614
15. Ewing, J., et al. (2020). Utilizing Hyperspectral remote sensing for soil gradation. *Remote Sensing, 12*(20), 3312. https://doi.org/10.3390/rs12203312
16. Fabre, S., Briottet, X., & Lesaignoux, A. (2015). Estimation of soil moisture content from the spectral reflectance of bare soils in the 0.4–2.5 μm domain. *Sensors (Switzerland)*. https://doi.org/10.3390/s150203262

17. FAO and ITU. (2018) *E-agriculture in action: Drones for agriculture, Journal of Chemical Information and Modeling*. Edited by G. Sylvester. Bangkok: FAO, ITU.
18. Fitzpatrick, A., Singhvi, A. & Arbabian, A. (2020). Spatial reconstruction of soil moisture content using non-contact thermoacoustic imaging. In *2020 IEEE SENSORS*. IEEE, pp. 1–4. https://doi.org/10.1109/SENSORS47125.2020.9278654.
19. Gago, J., et al. (2015). UAVs challenge to assess water stress for sustainable agriculture. *Agricultural Water Management, 153*, 9–19. https://doi.org/10.1016/j.agwat.2015.01.020
20. Gao, B. (1996). NDWI—A normalized difference water index for remote sensing of vegetation liquid water from space. *Remote Sensing of Environment, 58*(3), 257–266. https://doi.org/10.1016/S0034-4257(96)00067-3
21. Garcia Fernandez, M., et al. (2018). Synthetic aperture radar imaging system for landmine detection using a ground penetrating radar on board a unmanned aerial vehicle. *IEEE Access, 6*, 45100–45112. https://doi.org/10.1109/ACCESS.2018.2863572
22. Ge, X., et al. (2019). Combining UAV-based hyperspectral imagery and machine learning algorithms for soil moisture content monitoring. *Peer Journal, 7*, e6926. https://doi.org/10.7717/peerj.6926
23. Hardie, M. (2020). Review of novel and emerging proximal soil moisture sensors for use in agriculture. *Sensors, 20*(23), 6934. https://doi.org/10.3390/s20236934
24. Hassanalian, M., & Abdelkefi, A. (2017). Classifications, applications, and design challenges of drones: A review. *Progress in Aerospace Sciences, 91*, 99–131. https://doi.org/10.1016/j.paerosci.2017.04.003
25. Hassan-Esfahani, L., et al. (2015). Assessment of surface soil moisture using high-resolution multi-spectral imagery and artificial neural networks. *Remote Sensing, 7*(3), 2627–2646. https://doi.org/10.3390/rs70302627
26. Jabro, J. D., et al. (2014). Repeated freeze-thaw cycle effects on soil compaction in a clay loam in Northeastern Montana. *Soil Science Society of America Journal, 78*(3), 737–744. https://doi.org/10.2136/sssaj2013.07.0280
27. Jackisch, C., et al. (2020). Soil moisture and matric potential – An open field comparison of sensor systems. *Earth System Science Data, 12*(1), 683–697. https://doi.org/10.5194/essd-12-683-2020
28. Jia, Y. et al. (2015). GNSS reflectometry for remote sensing of soil moisture. In *2015 IEEE 1st international forum on Research and Technologies for Society and Industry, RTSI 2015 - Proceedings*. Institute of Electrical and Electronics Engineers Inc., pp. 498–501. https://doi.org/10.1109/RTSI.2015.7325147.
29. Kganyago, M., et al. (2020). Validation of sentinel-2 leaf area index (LAI) product derived from SNAP toolbox and its comparison with global LAI products in an African semi-arid agricultural landscape. *Remote Sensing Letters, 11*(10), 883–892. https://doi.org/10.1080/2150704X.2020.1767823
30. Klotzsche, A., et al. (2018). Measuring soil water content with ground penetrating radar: A decade of Progress. *Vadose Zone Journal, 17*(1), 180052. https://doi.org/10.2136/vzj2018.03.0052
31. Kornelsen, K. C., & Coulibaly, P. (2013). Advances in soil moisture retrieval from synthetic aperture radar and hydrological applications. *Journal of Hydrology, 476*, 460–489. https://doi.org/10.1016/j.jhydrol.2012.10.044
32. Liu, X., Dong, X., & Leskovar, D. I. (2016). Ground penetrating radar for underground sensing in agriculture: A review. *International Agrophysics, 30*(4), 533–543. https://doi.org/10.1515/intag-2016-0010
33. Lopez, C. D. & Giraldo, L. F. (2019). Optimization of Energy and Water Consumption on Crop Irrigation using UAVs via Path Design. In *2019 IEEE 4th Colombian Conference on Automatic Control (CCAC)*. IEEE, pp. 1–5. https://doi.org/10.1109/CCAC.2019.8921148.
34. Lu, B., et al. (2020). Recent advances of Hyperspectral imaging technology and applications in agriculture. *Remote Sensing, 12*(16), 2659. https://doi.org/10.3390/rs12162659

35. Matese, A., et al. (2015). Intercomparison of UAV, aircraft and satellite remote sensing platforms for precision viticulture. *Remote Sensing, 7*(3), 2971–2990. https://doi.org/10.3390/rs70302971
36. Messina, G., & Modica, G. (2020). Applications of UAV thermal imagery in precision agriculture: State of the art and future research outlook. *Remote Sensing, 12*(9), 1491. https://doi.org/10.3390/rs12091491
37. Novák, V., & Hlaváčiková, H. (2019). *Applied soil hydrology*. Springer International Publishing (Theory and Applications of Transport in Porous Media). https://doi.org/10.1007/978-3-030-01806-1
38. Rayburg, S., Thoms, M., & Neave, M. (2009). A comparison of digital elevation models generated from different data sources. *Geomorphology, 106*(3–4), 261–270. https://doi.org/10.1016/j.geomorph.2008.11.007
39. Ruosteenoja, K., Räisänen, J., & Pirinen, P. (2011). Projected changes in thermal seasons and the growing season in Finland. *International Journal of Climatology, 31*(10), 1473–1487. https://doi.org/10.1002/joc.2171
40. Salazar-Vazquez, J., & Mendez-Vazquez, A. (2020). A plug-and-play hyperspectral imaging sensor using low-cost equipment. *HardwareX, 7*, e00087. https://doi.org/10.1016/j.ohx.2019.e00087
41. Toivonen, M. E., Rajani, C., & Klami, A. (2021). Snapshot hyperspectral imaging using wide dilation networks. *Machine Vision and Applications, 32*(1), 1–11. https://doi.org/10.1007/s00138-020-01136-8
42. Uddin, M. A., et al. (2018). UAV-assisted dynamic clustering of wireless sensor networks for crop health monitoring. *Sensors, 18*(2), 555. https://doi.org/10.3390/s18020555
43. Van Looy, K., et al. (2017). Pedotransfer functions in earth system science: Challenges and perspectives. *Reviews of Geophysics, 55*(4), 1199–1256. https://doi.org/10.1002/2017RG000581
44. Wai-Lok Lai, W., Dérobert, X., & Annan, P. (2018). A review of ground penetrating radar application in civil engineering: A 30-year journey from locating and testing to imaging and diagnosis. *NDT & E International*. https://doi.org/10.1016/j.ndteint.2017.04.002
45. Wu, K., et al. (2019). A new drone-borne GPR for soil moisture mapping. *Remote Sensing of Environment, 235*, 111456. https://doi.org/10.1016/j.rse.2019.111456
46. Zajícová, K., & Chuman, T. (2019). Application of ground penetrating radar methods in soil studies: A review. *Geoderma, 343*, 116–129. https://doi.org/10.1016/j.geoderma.2019.02.024
47. Zhan, C., Zeng, Y., & Zhang, R. (2018). Energy-efficient data collection in UAV enabled wireless sensor network. *IEEE Wireless Communications Letters, 7*(3), 328–331. https://doi.org/10.1109/LWC.2017.2776922
48. Zheng, G., & Moskal, L. M. (2009). Retrieving Leaf Area Index (LAI) using remote sensing: Theories, methods and sensors. *Sensors, 9*(4), 2719–2745. https://doi.org/10.3390/s90402719

Ground-Penetrating Radar-Mounted Drones in Agriculture

Petri Linna, Antti Halla, and Nathaniel Narra

Abstract For precision farming, we need more and more accurate information not only about the crop, but also the soil. Surface measurement is fairly easy, with huge amounts of data being received from satellites all the time. With the help of drones, that data can still be refined, but the measurement price increases depending on the equipment as well as working time. With regard to soil measurement, measurement slows down and becomes more expensive.

The study mapped research papers of ground-penetrating radar and those different topics where they have been used. The topics were limited to agriculture only. The used frequencies were discovered from every topic. The study investigated artificial intelligence papers related to ground-penetrating radar and needs to begin an own artificial intelligence study in this subject. Finally, various concepts were evaluated for conducting ground-penetrating radar research. One of these concepts was to connect a ground-penetrating radar to a drone.

Keywords Soil compaction · Moisture content · Soil layers · Soil salinity · Drainage system · Roots · Biomass · Carbon · Robot · GPR

1 Background

When moving to the precision farming, increasingly exact information above ground as well as soil is needed. The first mentioned part is easy because there are a lot of data available from satellites all the time. With a drone, similar data can be obtained much more accurately, but depending on the equipment and spent working time it requires financial investment and data processing skills. Underground data collection is often time-consuming, expensive, destructive, labor-intensive, and point-based [34].

P. Linna (✉) · A. Halla · N. Narra
Tampere University, Pori, Finland
e-mail: petri.linna@tuni.fi; antti.halla@tuni.fi; nathaniel.narra@tuni.fi

T. Lipping et al. (eds.), *New Developments and Environmental Applications of Drones*,
https://doi.org/10.1007/978-3-030-77860-6_8

In Finland, soil variations can be very large for an individual field, so more measurement is needed to get a clearer picture of the field in question. In our previous MIKÄ DATA project, we made a grid at 50-m intervals for four different survey fields, giving each field about 30 to 40 soil sample points. The project collected data from several different sources, including satellite and drones acquired for the project. Drone imaging was performed every week on an area of 100 hectares, resulting in a complete time series for the entire growing season from two summers. The fields were scanned with electrical conductivity (EC) devices from Veris (Veris Technologies, Inc., USA) and Geocarta (Paris, France); the former also contain a Ph sensor and NIR camera. The EC-related devices provided a very accurate picture of the field and how the soil types in the field varied within the field and at different depths. The NIR camera also captured the soil mulch in the field. The Veris device also obtained a Ph value as point data about every twenty meters. Scans were performed on 12-m lines. The Veris Tech device reached a depth of one meter, and the Geocarta device reached a depth of two meters. This information was used to gain a better understanding about the soil and nutrient variation in the fields.

The same data were used to develop a yield forecast model. Historical data from the soil did not seem to play a major role in the development of the crop forecast model. The crop represents very strongly the current growing conditions of the field and the structure of the soil, so soil information does not provide new information during the harvest season for crop prediction [37, 38].

After the MIKÄ DATA project, the PeltoAI and BioEväät projects were started, which study in particular the growing conditions and water management of fields. The projects involve almost 20 fields. So far, field problems have been related to (1) water management for a variety of sectoral reasons (e.g., broken drainage or lack of water during the season), (2) low-organic matter, (3) compaction, (4) problematic elevation, (5) high pH values, (6) problematic history (e.g., old forest land, or food soil disposal), and (7) carbon emissions on peatlands. Some of these problems have also been identified from the available open data. For example, a broken drain was located from open orthophotos from the National Land Survey and moisture maps from the Sentinel satellite. Admittedly, the automatic detection of this problem would require the development of a more complex analysis. The surface elevation problems were found with the help of the openly available lidar data of the National Land Survey and the rain simulations made from them. NDVI maps collected from Sentinel satellite data or by drone provided a good indication of the problem areas in the field but did not directly identify the real problem accurately. The problem could be, among other things, compaction, water management, or a lack of organic matter.

The results of the above-mentioned projects strongly indicate that we need a solution that will allow us to find out more about soil compaction, moisture, soil layers, and the amount of biomass present. Based on this information, it would then be possible to assess the growth potential of the field more accurately and find solutions to the identified problem areas. Veris equipment was used in the Spring of 2018, but the measurements made did not focus on humidity and humidity calibration was not done. However, EC devices offer the possibility of moisture mapping in addition to other mapping [15, 24]. Ground-penetrating radar (GPR) is not technically a new invention. It has long been used, for example, in civil engineering and in the assessment of the condition of roads and railways. However, according to Liu et al., in a wide review paper, GPR is still a very little-studied technique in agriculture [34].

2 Introduction

This chapter focuses on GPR. First, the basic features of GPR are introduced. GPR is used, for example, to exam the structures of concrete or find objects in the soil. So GPR can provide information without having to break up the structure. The GPR devices use microwave bands of UHF and VHF, which usually means in practice about 100 MHz to 2 Ghz. Usually, the device can use only a certain frequency band, thus concentrating only on the identification of certain matters. The low frequencies will go to several ten meters in the soil, but their precision weakens according to depth. Higher frequencies separate more details but with them get only to the surface layers. The GPR device transmits the signal and the same device contains a receiver with which it catches the signal. From the force, speed, and echo of the signal echo, the device is able to separate matters from the target.

The operating principle of GPR is illustrated in Fig. 1. It transmits the signal (Tx), and the signal will come back from different layers or targets as a reflected wave. This signal is received with Rx, and then signal is processed and showed to the user.

In the literature review were mainly used the sources of IEEE and Springer. Because there were a lot of different sub-topics of applications of GPR, only topics that are linked or beneficial to the agriculture were focused. The papers that are

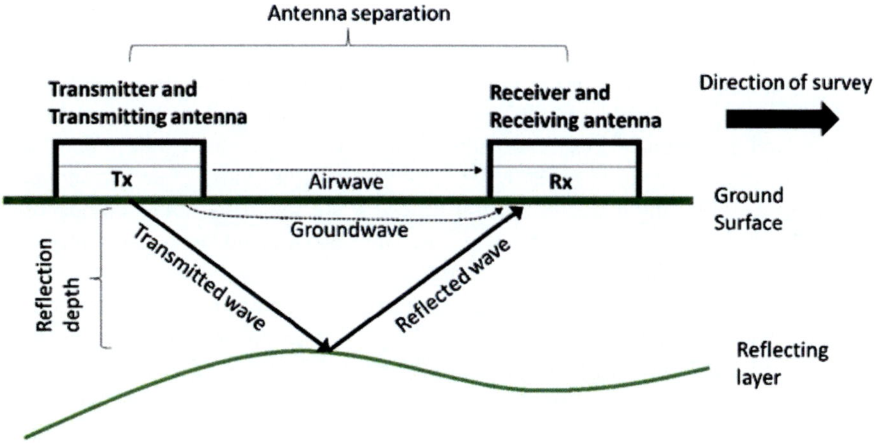

Fig. 1 Structure of GPR

related to the AI have been filtered according to the age of paper, because of quick developments of AI during the ten last years.

In the last chapter has been utilized own experience of different concepts. The robot concept has been studied earlier, and commercial site equipment has shortly tested. The drone equipment has been used for couple of years, so it helps to build GPR and drone concept. The combination of the trailer and the all-terrain vehicle of a study has also been analyzed at a practical level.

The research questions relate to the mapping of GPR utilization potential in the pilot fields. (1) What issues can be identified with GPR in agricultural perspective? (2) What the frequencies of GPR should be used in agriculture? (3) How to utilize AI in GPR data analysis? (4) What kind of concept should build where is utilized drone with GPR?

3 What Issues Can Be Identified with Ground Radar?

The following subsections cover things that can be determined from the soil with GPR. The aim is to find GPR studies from agriculture sector or which can apply to agriculture sector.

3.1 Soil Compaction

Soil compaction is a big problem in fields because it prevents the roots from growing properly and prevents water from being stored in the soil. Big and heavy machines are constantly causing fresh compaction. Certain soil types are also more sensitive to compaction. In Finland, the challenge for beet cultivation is also that it is collected in late autumn with large machines, which means that the soil is usually very wet and easily becomes compacted.

According to the results of [1], compaction causes many problems in cultivation, as strong compaction reduces the macropores within the soil matrix and hence reduces water content. These cause degradation to the soil and thus reduce the amount of crop. They used an 800 MHz antenna in their study. They had induced compaction by driving a tractor several times on the same lines. The compaction caused by the tractor was detectable by GPR. They also noticed how soil compaction restricted root growth.

Wang et al. [51] used a 500 MHz antenna. They noticed strong correlation between electromagnetic wave velocity and bulk density. In addition, they noticed that the soil dielectric constant was strongly dependent on the soil water content. They estimated that a plenty of field research would still be required to assess the relationship between soil compaction and GPR.

3.2 Moisture Content

Moisture measurement is strongly associated with soil dielectric permittivity, the dielectric constant, and electrical conductivity. Moisture can be calculated using these values, but in addition, the frequency of the measuring device must be taken into account [7, 8]. Dielectric permittivity means the ability of a medium to allow passage of electromagnetic energy, and electrical conductivity means the availability and mobility of electrical charges in the medium [48].

Lu et al. [36] evaluated that soil moisture content could be measured in three different ways based on their scale: point (e.g., the gravimetric method), intermediate (e.g., GPR), and coarser scale (e.g., satellite). According to their research, GPR can achieve high measurement accuracy, but more research is needed with different soil types and GPR methods. They used 250 MHz frequencies.

There have been efforts to measure moisture from satellites in "The Soil Moisture and Ocean Salinity" mission and thus obtain moisture data from large areas, but the problem is that the moisture is only above the ground or a few centimeters from the ground. For this reason, a device such as GPR is needed, whereby moisture can be measured to a greater depth [13]. Algeo et al. [2] tested three different frequencies of GPR: 250, 400, and 1000 MHz, which was used for the evaluation of early time signals and signal traveling time. The results look promising but still need confirmation before commercial implementation.

In many studies, it has been noticed that it is possible to measure the volume content of moisture in the soil [26, 40, 42, 53]. Wu et al. [54] recently conducted a study where they built a drone with GPR. Over the 700 MHz frequency, they faced the challenge of soil roughness associated with Rayleigh's criterion. They experienced disturbances in the 400–500 MHz range, so the final measurement range chosen was 500–700 MHz. The study yielded promising results but still needs further research and hardware adjustment. Liu et al. [33] obtained the best result with the frequency of 900 MHz, and during their research, they made moisture and root measurements to obtain more accurate results. Benedetto [9] used 600 and 1600 MHz antennas in their research.

Despite several recent studies, it is not yet entirely clear which frequencies are suitable. Several of the above-mentioned studies state that the results are promising, but the functionality of the method still requires confirmation.

3.3 Soil Layers

GPR provides a good indication of the lithological soil layers of the soil [42]. Roads and railways in particular are interested in the condition and depths of the soil layers. In fields, research can focus on the identification of soil layers or the identification of the point between different layers. Benedetto and Tosti [10] made an evaluation of clay content and were able to identify even just a couple of percentages of clay content. Their findings mainly focused on the 570–630 MHz range.

The Geological Survey of Finland (GTK) has carried out large geological surveys in Finland. They have for many years researched more than 400 peat production areas by GPR, utilizing 100, 200, and 250 MHz antennas. In the last few years, they have also developed and used UAVs (i.e., drones) to collect multispectral data to support GPR data [46]. The soil type of some of Finland's fields is peat, so this research also provides a really broad and good basis for field research.

3.4 Soil Salinity

Soil salinity in Europe is not a major issue. However, in southern coastal areas of Europe, this is starting to become a problem when seawater mixes with groundwater. Daliakopoulos et al. [16] state in their research that remote sensing and especially satellite data are a promising measurement method for soil salinity. There is still a need to develop measurement of soil salinity to gain a deeper understanding of the soil and develop soil protection and remediation as a whole.

Electrical conductivity represents a material's ability to conduct electric current, and it also indicates the amount of mineral salts it contains. A high-conductivity level indicates high salinity. In fields, the amount of salts is increased by fertilization. In Finland, the fertility analysis of soil sample conductivity is usually measured as electrical conductivity in mS/cm and is reported as a leading number by multiplying the obtained measurement result by ten. Usually, the leading number is less than 2.5. Higher conductivity means that plants have more difficulty to get nutrients and water.

According to many studies, the soil salinity can be measured by GPR [17, 42, 55]. Soil salinity has a very strong link to soil electrical conductivity, so soil salinity can be calculated based on the electrical conductivity and soil moisture. Awak et al. [6] used 200 MHz, and their research result was that GPR can be used to get even high-resolution electrical conductivity maps.

3.5 Drainage System

Detection of various objects is the easiest area for GPR. In agriculture, this may relate to the examination of locations of drainage pipes, if the exact location of the drains is not known. GPR detects easily, especially older clay-made drainers, because they will give a stronger echo, instead of new plastics, which have a weaker double echo from the top and inside of the pipe. The best echo from drainpipes by GPR will be obtained by driving over them vertically.

As a result of our own GPR testing, the pipes can be identified well at 1000 MHz but also at lower frequencies. The frequency is linked to how deep the pipes are. In our own field tests, 1000 MHz was sufficient for a depth of up to 3–4 m, so the frequency was suitable for checking drains. In Finland, the need to search for a drainage pipe system is quite small because of a good centralized long-term archiving procedure of drainage pipe maps. Allred [4] obtained good results with 250 MHz antennas. However, in a subsequent study, Allred et al. [3] found that drainage pipes are easier to recognize from homogeneous rather than heterogeneous soils.

3.6 *Roots, Biomass, and Carbon*

The measurement of root systems using GPR has been studied mainly for trees, achieving 85% accuracy [42]. The diameter of a tree's fine roots is less than 2 mm and that of finest roots smaller than 1 mm [25]. When measuring root systems in fields, the challenge is that the root sizes are much smaller. Winter wheat has fine roots of 0.5–2.0 mm in diameter for example and very fine roots that are less than 0.5 mm in diameter [35]. Root measurement can be focused on many different things: diameter, biomass, architecture, depth, quantity, and detection [34]. The importance of root system measurement is related to the rising trend of measuring the amount of biomass in a field, both underground and above ground. The amount of carbon formed from this biomass can be estimated, which in turn is strongly linked to the condition of the field and to environmental issues such as the evaporation of carbon dioxide from the field.

Shen et al. [44] quantified soil carbon with GPR, and the best detection for root diameter was 56% when they used a GPR frequency of 1800 MHz. Liu et al. [35] noticed that accuracy was linked to soil moisture and different soil conditions, with wet clay soil being better than dry sandy soils.

The study of crop root measurement using GPR is still very limited. Research is required to find out the impacts of other soil factors that can affect the GPR signal and accuracy of root estimations. GPR is a significant potential technique in estimating the ratio of surface biomass to underground biomass for different varieties. This data could be used to make prediction models that would later be calculated for biomass calculated only from the surface.

4 What Frequencies Should Be Used in Agriculture?

The aim is to find out which frequencies have been used in the applications of the first research question. The previous section listed issues that can be clarified with GPR and the frequencies used in those cases. The lowest frequencies started at about 100 MHz and the highest was closer to 2 GHz.

Liu et al. [34] have extensively explored potential issues from an agricultural perspective as well as their frequencies. Tables 1 and 2 summarize that the best

Table 1 Soil measurement frequencies of GPR, edited from Liu et al. [34]

Target	Frequency
Soil clay content	500, 600
Hardpan	500
Soil moisture, soil water	400, 100, 200–800, 600, 1600

Table 2 Root measurement frequencies of GPR, edited from Liu et al. [34]

Target	Frequency
Root diameter and biomass	500, 800
Root biomass and root architecture	1000
Root diameter	900, 1000
Root biomass	1500
Root diameter and root depth	500, 900, and 2000
Root detection	900
Root number	900

frequencies are somewhere around 500 and 1000 MHz. On the other hand, it is difficult to give any frequency that would fit every case. Some studies have used devices that have a wider frequency range instead of just one frequency. Especially, fine roots require a much higher frequency than others.

It would also be possible to try the gprMax tool [52], but it has only three different antennas, which limits the matching of suitable frequencies. GPR still needs a lot of research, especially in the agricultural sector, to get a clear picture of the usability of different frequencies for finding different features. In addition, there are many things to be investigated, and it is not yet possible to establish with certainty the connections between all of them.

5 AI Development

In the previous research paper, we mapped out how much AI had been utilized in the use of GPR data and what had been the identifiable issues during the past decade. That study found more than 20 papers [32].

According to the summary of Table 3, the biggest efforts have been made to identify landmines and various other objects. Other issues were also explored in the individual papers. From an agricultural perspective, the most interesting are such as "Water content and its prediction," "Layer properties: Density, permittivity," "Moisture content," and "Compaction." Research must be continued in this area, and efforts must be made to expand it to find more AI studies that focus on what is necessary from an agricultural point of view. Many of the studies mentioned in the table had been performed based on simulated data generated by the gprMax tool. In this simulation tool, it is possible to use three different antennas, which were

Table 3 GPR and AI research areas

Research area	Article
Layer properties: density and permittivity	[11]
Cover depth and diameter of rebar	[22]
Moisture content	[22]
Corrosion, infrastructure monitoring	[5]
Bridge deck evaluation (density, moisture, rebar locations, corrosion)	[28]
Detection of hidden surface crevasses on glaciers	[53]
Landmines	[23, 31, 40, 41, 49]
Water content and its prediction	[57]
Small targets, holographic subsurface penetration	[56]
Debonding, pavement monitoring	[47]
Pipes, cables, and course of their track	[43]
Deep and shallow crevasses	[50]
Objects (generally)	[14, 19, 29, 39, 45, 58]

1.5, 1.2, and 0.4 GHz. However, the study did not focus on identifying the GPR frequencies [32].

Utilization of AI and especially the utilization of neural networks with newer technologies require more research in order to interpret GPR data more automatically. The aim is to explore the potential of AI for GPR data in agricultural applications. Gathering GPR data the traditional way (walking with instruments) can be slow, and clarifying the ground truth is challenging, especially when one needs to go deeper into the soil.

The following aspects need to be solved before practical AI development:

1. Training data:
   (a) How to put labels to the data?
   (b) Ground truth: human evaluation vs. real knowledge?
2. Data:
   (a) Real vs. simulated
   (b) Before vs. after season or during growing season
   (c) Own vs. open data
   (d) Once vs. time series data
   (e) Real vs. post-analyses
3. Finding a suitable AI technique
   (a) Neural networks vs. machine learning
   (b) Existing analyses [18]

AI needs a lot of training data. In GPR data, one challenge is to develop labeling methods for marking drainage pipes for example. In addition, it is becoming a challenge to find out the ground truth of the soil. Many traditional soil samples would probably be needed to confirm the entire data set and calibrate the results with them.

Obtaining data for the development of AI requires human and equipment resources. Collecting data may also involve challenges related to the time of the collecting period, i.e., weather conditions need to collect data series, or seasonal growing requirements. Real data should be used in AI development so that the AI model can be trained correctly, but it may be possible to use simulated data to test different initial models. When collecting data from fields, the challenges of data transparency are easily encountered. However, some open data sets can be found in foreign sources (example sources [20, 30], Data.gov,[1] RGPR,[2] Data.world[3]). This problem was encountered in the study conducted by [32]. In many of these studies, researchers used simulated data, which had been made with gprMax.

Choosing the right kind of AI technology requires the capture and tagging of the appropriate data. The AI techniques used were listed in our previous paper, so this requires more practical level testing. AI has been used quite little in the study of GPR data, as only around twenty papers were found [32]. However, it is also good to note the numerous existing methods of analysis.

6 The Concept Model of a Drone with GPR

We have outlined up to four different concepts for implementing GPR measurements (Fig. 2). In our previous study, we introduced concept of an autonomous robot mounted with a GPR device [32]. The robot designed for logistical purposes is based on the ROS platform; however, its design will also consider the needs of a GPR, such as power supply, navigation accuracy, and bigger tires for improved mobility. An autonomous robot can help with improving time and labor required for data collection. This is especially convenient if there is a need to conduct time series data collection in the same area. The constraint is the power supply needs of the robot and the challenges associated with the terrain, such as a soft or watery field. The use of the robot is limited to pre-sowing or post-harvest periods of the season. Field tests have not yet been performed using a robot.

The all-terrain vehicle (ATV) has been utilized in previous field scans carried out by Geocarta[4] in the MIKÄ DATA project. The ATV proved to be a really fast and efficient data collection concept. In that case, the measuring equipment was placed in self-made trailer. Neither the shape of the terrain nor the humidity caused measurement problems. A trailer may also be required in GPR case because it allows

[1]https://catalog.data.gov/dataset?tags=gpr.

[2]https://emanuelhuber.github.io/RGPR/80_RGPR_GPR-data-free-to-download/.

[3]https://data.world/datasets/gpr.

[4]https://geocarta.net/.

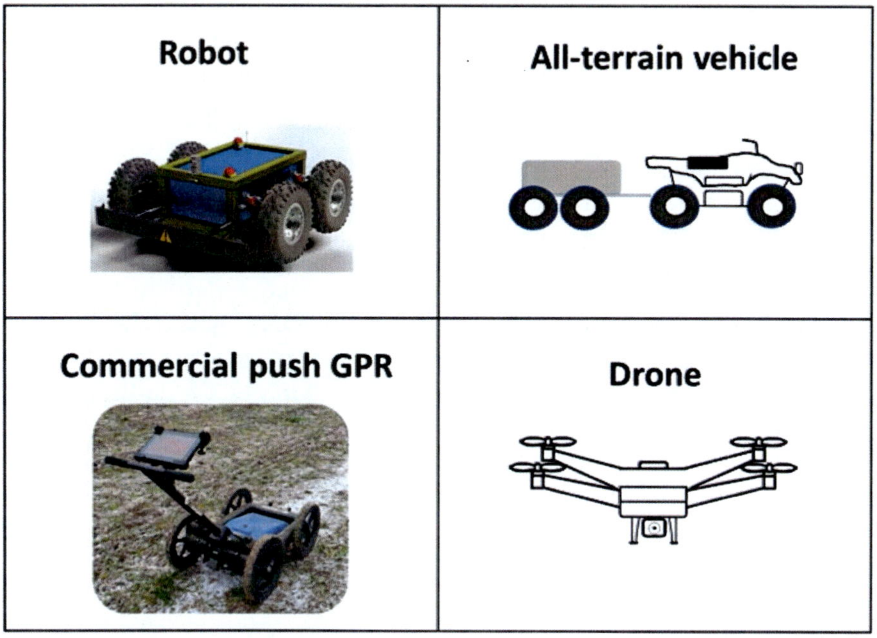

Fig. 2 Different platforms for conducting GPR surveys

the simultaneous use of other measuring devices. For GPR use only, the use of an ATV may be slightly oversized. A significant advantage of an ATV is that it provides the measuring equipment with enough power for longer periods of measurements. An ATV requires a dedicated navigation system to follow a pre-calculated route.

Commercial GPR products provide a quick way to carry out measurements. At the end of 2020, the Quantum Imager (USRadar Inc., USA) Triple Frequency[5] (100, 250, and 1000 MHz) was tested with assistance from Kontram (Kontram Oy,[6] Finland). The Quantum allowed up to three different frequencies to be used same time, which is still exceptional. In this way, preliminary field results and insights into the right type of GPR were obtained. The equipment was used to collect data for three fields, which means over ten kilometers walked routes in the fields. The practical experience was interesting and very necessary to develop the concepts and deepen the GPR skills. The challenge was to plan the exact route in advance, which in practice had to be done with a separate device and control the route based on it. It is also a constraint to carry out larger measurements, as the measurement is quite laborious if there is even a little vegetation in the field that limits the movement of the wheels. Vegetation is also challenge for the robot concept. In this case, it became

[5]https://usradar.com/quantum-imager-triple-frequency-gpr-system/.

[6]https://www.kontram.fi/en/etusivu.html.

clear that tires size and softness are important. Even small irregularities cause quite a bit of equipment jumping, as the soil froze this worsened. So there is needed wide tires with low air pressure. Finding a suitable weather became a challenge in these measurements. There were quite rainy weeks at the time of the quote, so this should be noticed in all measurement methods.

The aim of this study is to design the concept of a combination of a GPR device and a drone, which is suitable for agriculture needs. This research revealed three studies where a combination of GPR equipment and a drone had also been built. Wu et al. [54] focused on soil moisture mapping in the 500–700 MHz frequency range. Their hardware configuration consisted of a drone, a lightweight vector network analyzer (Planar R60), a transmitting and receiving (monostatic) antennas, a power bank, and a smartphone. The results were promising but require further investigation, such as antenna calibration on different surfaces. García-Fernández et al. [21] approached the use of drone and GPR from a slightly different perspective and developed a portable and easily deployable air-launched GPR scanner. The hardware included a radar module whose range was 100 MHz–6 GHz, antenna, GNNS, RTK, micro-computer (Raspberry Pi), and batteries. The third finding was the study by Cerquera et al. [12] where they used a software-defined radio (SDR) platform as GPR. They installed an SDR platform on the drone using a USRP B210 device[7] by Ettus (Ettus Research, Texas, USA), which can programmatically select the frequencies to be used between 70 MHz and 6 GHz. Their research focused on identifying landmines. The GPR hardware thus consisted of a USRP platform and antennas connected to it. The drone also had an optical sensor for altitude control.

Commercial integrations of drones and GPR are available on the market. Novatest (Novatest S.r.l., Ancona, Italy) offers a dual configuration setup[8] where, for example, the DJI Matrix 600 Pro (DJI, Shenzhen, China) is used as the drone platform and the ground probe is alternatively a CBD GPR 200/400/800 MHz or low-frequency GPR plug-in SUBECHO model (RadarTeam AB, Boden, Sweden) using 70/150 MHz. However, depending on the model and battery, the promised flight time is only 15–20 min, which in practice means about two kilometers of measuring distance. At lower frequencies, the size of the antenna poses its own challenge, because the length of the antenna increases rapidly as the frequency decreases.

The advantage of a drone is to carry out data collection during the growing season and ease of collection from a rough surface too, e.g., after plowing. In addition, the drone has the advantage of high speed, although the challenges are a relatively short flight time and the weather-related flight constraints. Furthermore, flight permission might be challenging for some fields due to regulatory constraints. According to this study, the frequencies required are mainly between 500 and 1000 MHz with even higher frequencies required for root systems.

[7] https://www.ettus.com/all-products/ub210-kit/.

[8] https://www.novatest.it/en/civil-engineerings/uav-gpr-cobra/.

Fig. 3 USRP B210

Based on the above findings, the drone concept will consist of the following devices. The DJI Matrix 600 Pro is a potential choice, due to its carrying capacity, commercial availability, and as well as experiences from other studies. The GPR will be based on the USRP device (Fig. 3), as it is an open development platform and it allows to choose the frequencies from wide range. In addition, it allows the use of several frequencies by varying them during the measurement, so that measurement data of several frequencies are obtained from almost the same measurement point. A significant advantage of the USRP device is its relative low cost. A major challenge in the USRP is that it requires considerable in-depth knowledge of software coding and signal processing. When building the drone concept, it should also be taken into account that the USRP device can be used for other concepts. The drone must be equipped with a precision GNSS system in order to fly at a fairly low altitude and on the route planned. In previous studies, the flights have taken place at a higher altitude of 100–150 m, which has made it possible to combine the images into one image afterward, and in addition, there have been fixed points in the fields for alignment.

Measurement is affected by the permeability and conductivity of the soil. According to the International Telecommunication Union, the moisture content is the major factor determining permittivity and conductivity [27]. Therefore, it is necessary to make moisture measurements in the fields to be measured to calibrate the results.

In the PeltoAI project, currently Tomst TMS-4[9] (Tomst s.r.o., Prague, Czech Rep.) soil moisture sensors are installed in seven test fields, at three different locations in each field. They are set up so that two sensors measure the problem area of the field and one measures the average area grown according to the NVDI value obtained from the satellite, which serves as a benchmark for the worse points. In addition, some measuring points also have another sensor that is 40 cm deep.

One of the test fields has been scanned with a Veris device, providing a map of the field conductivity. If necessary, calibration measurements can also be taken from the field with a separate soil compaction meter. At least three soil samples have been taken from each test field to provide accurate information on the soil type of the soil and the conductivity at the time of measurement. The real-time moisture and

[9] https://tomst.com/web/en/systems/tms/tms-4/.

electrical conductivity measurement sensors (LSE01,[10] Dragino Technology Co., Ltd., Shenzhen, China) will be ordered for the season 2021.

The soil density meter, soil samples, moisture, and conductivity sensors provide ground truth values to calibrate and compare to the GPR data. These are very important steps to enable the development of AI with validated data.

7 Conclusions and Discussion

This chapter explored the factors and information about soil that can be determined by GPR. The most commonly used frequency bands would appear to be concentrated in the 500–1000 MHz range although much higher frequency bands were used for the study of root systems. The lower frequencies would no longer seem to provide any added value in the study of fields, as factors affecting the growth conditions of a field are at a depth of about 0–2 m. In root research, the challenge is especially the size of the finest roots, and their measurement seems to be challenging and needs more research. In the AI section, the issues that take priority were researched. Gathering data and making training material will involve a major research effort before the actual development of AI can be done.

There were introduced four GPR concepts. The concept of using drones with GPR was investigated more deeply. The advantage of a drone is especially fast implementation if there is a need to study large areas of land, as well as the ability to measure a rough surface target, and the ability to measure during the growing season. The challenge would be to get the device close enough to the ground if higher frequency GPR is used, which needs to be closer to the soil. At lower frequencies, this is less of a problem. However, the accurate maximum flight altitudes of the GPR frequencies have yet to be determined.

The next goal is to study the implementation of the necessary analyses. Each area to be studied might require a different method of analysis. It must be ascertained whether some analysis requires certain specific frequency ranges or two different frequencies, or whether the analyses are flexible in terms of frequencies. Another goal is to study USRP platform and implement it in drone. The third goal is to define the possibility of using AI in data analysis.

This study focused on GPR and its potentialities in agriculture. Also AI development needs were evaluated. Last were introduced four concepts and specially of drone with GPR. In several of our projects, we have tested different measurement technologies, and based on them, we came to the same conclusions as [48] that in the future there will be a combination of many different sensors. This is due to the fact that each device has its own advantages and disadvantages.

---

[10] https://www.dragino.com/products/lora-lorawan-end-node/item/159-lse01.html.

Acknowledgments MIKÄ DATA[11] project has been funded from EU's The Agricultural European Innovation Partnership (EIP-AGRI) program. PeltoAI and BioEväät projects[12] are funded from: European Regional Development Fund and European Agricultural Fund for Rural Development. The authors grateful for assistance from Kontram Oy in testing with their Quantum Image Triple Frequency GPR device.

References

1. Akinsunmade, A., Tomecka-Suchoń, S., & Pysz, P. (2019). Correlation between agrotechnical properties of selected soil types and corresponding GPR response. *Acta Geophysica, 67*(6), 1913–1919.
2. Algeo, J., Slater, L., Binley, A., Van Dam, R. L., & Watts, C. (2018). A comparison of ground-penetrating radar early-time signal approaches for mapping changes in shallow soil water content. *Vadose Zone Journal, 17*(1), 180001.
3. Allred, B., Wishart, D., Martinez, L., Schomberg, H., Mirsky, S., Meyers, G., Elliott, J., & Charyton, C. (2018). Delineation of agricultural drainage pipe patterns using ground penetrating radar integrated with a real-time kinematic global navigation satellite system. *Agriculture (Switzerland), 8*(11), 167.
4. Allred, B. J. (2013). A GPR agricultural drainage pipe detection case study: Effects of antenna orientation relative to drainage pipe directional trend. *Journal of Environmental and Engineering Geophysics, 18*(1), 55–69.
5. Alvarez, J. K., & Kodagoda, S. (2018). Application of deep learning image-to-image transformation networks to GPR radargrams for sub-surface imaging in infrastructure monitoring. In *Proceedings of the 13th IEEE Conference on Industrial Electronics and Applications, ICIEA 2018* (pp. 611–616).
6. Awak, E., George, A., Urang, J., & Udoh, J. (2017). Determination of soil electrical conductivity using ground penetrating radar (GPR) for precision agriculture. *International Journal of Scientific & Engineering Research, 8*(1).
7. Behari, J. (2005a). *Dielectric Constant of Soil* (pp. 92–106). Dordrecht: Springer.
8. Behari, J. (2005b). *Soil Moisture Models* (pp. 107–124). Dordrecht: Springer.
9. Benedetto, A. (2010). Water content evaluation in unsaturated soil using GPR signal analysis in the frequency domain. *Journal of Applied Geophysics, 71*(1), 26–35.
10. Benedetto, F., & Tosti, F. (2013). GPR spectral analysis for clay content evaluation by the frequency shift method. *Journal of Applied Geophysics, 97*, 89–96.
11. Caorsi, S., & Stasolla, M. (2009). A Machine Learning Algorithm for GPR Sub-surface Prospection. In *2009 Mediterrannean Microwave Symposium (MMS)*.
12. Cerquera, M. R. P., Montaño, J. D. C., & Mondragón, I. (2017). UAV for landmine detection using SDR-based GPR technology. In *Robots Operating in Hazardous Environments*. IntechOpen.
13. Chantasen, N., Boonpoonga, A., Athikulwongse, K., Kaemarungsi, K., & Akkaraekthalin, P. (2020). Mapping the physical and dielectric properties of layered soil using short-time matrix pencil method-based ground-penetrating radar. *IEEE Access, 8*, 105610–105621.
14. Chen, X. L., Tian, M., & Yao, W. B. (2005). GPR signals de-noising by using wavelet networks. In *2005 International Conference on Machine Learning and Cybernetics, ICMLC 2005* (pp. 4690–4693).

[11]https://www.avoinsatakunta.fi/mikadataeng/.

[12]https://www.avoinsatakunta.fi/peltoaibioevaat_eng/.

15. Corwin, D. L., & Lesch, S. M. (2003). Application of soil electrical conductivity to precision agriculture. *Agronomy Journal, 95*(3), 455–471.
16. Daliakopoulos, I. N., Tsanis, I. K., Koutroulis, A., Kourgialas, N. N., Varouchakis, A. E., Karatzas, G. P., & Ritsema, C. J. (2016). The threat of soil salinity: A European scale review. *Science of The Total Environment, 573*, 727–739
17. Doolittle, J. A., Jenkinson, B., Hopkins, D., Ulmer, M., & Tuttle, W. (2006). Hydropedological investigations with ground-penetrating radar (GPR): Estimating water-table depths and local ground-water flow pattern in areas of coarse-textured soils. In *Geoderma*, vol. 131 (pp. 317–329). Amsterdam: Elsevier.
18. Economou, N., Vafidis, A., Benedetto, F., & Alani, A. M. (2015). GPR data processing techniques. In A. Benedetto & L. Pajewski (Eds.), *Civil Engineering Applications of Ground Penetrating Radar* (pp. 281–297). Cham: Springer.
19. Elsaadouny, M., Barowski, J., & Rolfes, I. (2019). The subsurface objects classification using a convolutional neural network. In *2019 IEEE 10th Annual Information Technology, Electronics and Mobile Communication Conference, IEMCON 2019* (pp. 874–877).
20. Florez-Lozano, J., Caraffini, F., Gongora, M., & Parra, C. (2019). Ground penetrating radar dataset. Retrieved from https://figshare.dmu.ac.uk/articles/dataset/Ground_penetrating_radar_dataset/8323049
21. García-Fernández, M., Álvarez López, Y., De Mitri, A., Castrillo Martínez, D., Álvarez-Narciandi, G., & Las-Heras Andrés, F. (2020). Portable and easily-deployable air-launched GPR scanner. *Remote Sensing, 12*(11), 1833.
22. Giannakis, I., Giannopoulos, A., & Warren, C. (2019). A machine learning-based fast-forward solver for ground penetrating radar with application to full-waveform inversion. *IEEE Transactions on Geoscience and Remote Sensing, 57*(7), 4417–4426.
23. Giovanneschi, F., Mishra, K. V., Gonzalez-Huici, M. A., Eldar, Y. C., & Ender, J. H. G. (2019). Dictionary learning for adaptive GPR landmine classification. *IEEE Transactions on Geoscience and Remote Sensing, 57*(12), 10036–10055.
24. Hawkins, E., Fulton, J., & Port, K. (2017). Using Soil Electrical Conductivity (EC) to Delineate Field Variation. Retrieved from https://ohioline.osu.edu/factsheet/fabe-565
25. Helmisaari, H.-S., Lehto, T., & Makkonen, K. (2000). Fine roots and soil properties. In E. Mälkönen (Ed.), *Forest Condition in a Changing Environment: The Finnish Case* (pp. 203–217). Dordrecht: Springer.
26. Huisman, J., Hubbard, S., Redman, J., & Annan, P. (2003). Measuring soil water content with ground penetrating radar: A review. *Vadose Zone Journal, 2*, 476–491.
27. ITU-R. (2017). P.527: Electrical characteristics of the surface of the earth. Retrieved from https://www.itu.int/rec/R-REC-P.527-4-201706-I/en
28. Kaur, P., Dana, K. J., Romero, F. A., & Gucunski, N. (2016). Automated GPR rebar analysis for robotic bridge deck evaluation. *IEEE Transactions on Cybernetics, 46*(10), 2265–2276.
29. Kim, N., Kim, S., An, Y. K., & Lee, J. J. (2019). Triplanar imaging of 3-D GPR data for deep-learning-based underground object detection. *IEEE Journal of Selected Topics in Applied Earth Observations and Remote Sensing, 12*(11), 4446–4456.
30. Koyan, P., & Tronicke, J. (2020). 3d modeling of ground-penetrating radar data across a realistic sedimentary model. *Computers & Geosciences, 137*, 104422.
31. Lameri, S., Lombardi, F., Bestagini, P., Lualdi, M., & Tubaro, S. (2017). Landmine detection from GPR data using convolutional neural networks. In *25th European Signal Processing Conference, EUSIPCO 2017* (vol. 2017, pp. 508–512).
32. Linna, P., Aaltonen, T., Halla, A., Grönman, J., & Narra, N. (2020). Conceptual design of an autonomous rover with ground penetrating radar: Application in characterizing soils using deep learning. In *2020 43rd International Convention on Information, Communication and Electronic Technology (MIPRO)* (pp. 1134–1139)
33. Liu, X., Chen, J., Butnor, J. R., Qin, G., Cui, X., Fan, B., Lin, H., & Guo, L. (2020). Noninvasive 2D and 3D mapping of root zone soil moisture through the detection of coarse roots with ground-penetrating radar. *Water Resources Research, 56*(5), e2019WR026930.

34. Liu, X., Dong, X., & Leskovar, D. I. (2016). Ground penetrating radar for underground sensing in agriculture: A review. International Agrophysics, 30, 533–543.
35. Liu, X., Dong, X., Xue, Q., Leskovar, D. I., Jifon, J., Butnor, J. R., & Marek, T. (2018). Ground penetrating radar (GPR) detects fine roots of agricultural crops in the field. *Plant Soil, 423*, 517–531.
36. Lu, Y., Song, W., Lu, J., Wang, X., & Tan, Y. (2017). An examination of soil moisture estimation using ground penetrating radar in desert steppe. *Water, 9*(7), 1–11. https://doi.org/10.3390/w9070521
37. Nevavuori, P., Narra, N., Linna, P., & Lipping, T. (2020). Crop yield prediction using multitemporal UAV data and spatio-temporal deep learning models. *Remote Sensing, 12*(23), 1–18. https://doi.org/10.3390/rs12234000
38. Nevavuori, P., Narra, N., & Lipping, T. (2019). Crop yield prediction with deep convolutional neural networks. *Computers and Electronics in Agriculture, 163*, 104859.
39. Pham, M., & Lefèvre, S. (2018). Buried object detection from B-scan ground penetrating radar data using faster-RCNN. In *IGARSS 2018–2018 IEEE International Geoscience and Remote Sensing Symposium* (pp. 6804–6807).
40. Picetti, F., Testa, G., Lombardi, F., Bestagini, P., Lualdi, M., & Tubaro, S. (2018). Convolutional autoencoder for landmine detection on GPR Scans. In *2018 41st International Conference on Telecommunications and Signal Processing, TSP 2018*. Piscataway: Institute of Electrical and Electronics Engineers Inc.
41. Reichman, D., Collins, L. M., & Malof, J. M. (2017). Some good practices for applying convolutional neural networks to buried threat detection in Ground Penetrating Radar. In *2017 9th International Workshop on Advanced Ground Penetrating Radar, IWAGPR 2017 - Proceedings*. Piscataway: Institute of Electrical and Electronics Engineers Inc.
42. Ristic, A., Petrovacki, D., & Vrtunski, M. (2014). Ground penetrating radar technology - the usage in agriculture. *Research Journal of Agricultural Science, 46*, 53–58. ISSN: 2066-1843
43. Seyfried, D., Busche, A., Janning, R., Schmidt-Thieme, L., & Schoebel, J. (2012). Information extraction from ultrawideband ground penetrating radar data: A machine learning approach. In *2012 the 7th German Microwave Conference, GeMiC 2012*.
44. Shen, X., Foster, T., Baldi, H., Dobreva, I., Burson, B., Hays, D., Tabien, R., & Jessup, R. (2019). Quantification of soil organic carbon in biochar-amended soil using ground penetrating radar (GPR). *Remote Sensing, 11*(23), 1–12. https://doi.org/10.3390/rs11232874
45. Sonoda, J., & Kimoto, T. (2019). Object identification form GPR images by deep learning. In *Asia-Pacific Microwave Conference Proceedings, APMC* (vol. 2018, pp. 1298–1300).
46. Sutinen, H. (2020). Service package with a ground-penetrating radar and a drone for peat research. Retrieved from https://www.gtk.fi/en/service-package-with-a-ground-penetrating-radar-and-a-drone-for-peat-research/
47. Todkar, S. S., Le Bastard, C., Ihamouten, A., Baltazart, V., Dérobert, X., Fauchard, C., Guilbert, D., & Bosc, F. (2017). Detection of debondings with ground penetrating radar using a machine learning method. In *2017 9th International Workshop on Advanced Ground Penetrating Radar, IWAGPR 2017 - Proceedings* (vol. 2017). Institute of Electrical and Electronics Engineers Inc.
48. Van Meirvenne, M. (2015). Advanced electric and electromagnetic methods for the characterization of soil. In A. Benedetto & L. Pajewski (Eds.), *Civil Engineering Applications of Ground Penetrating Radar* (pp. 343–359). Cham: Springer.
49. Vu, T. H., Nguyen, L., Guo, T., & Monga, V. (2018). Deep network for simultaneous decomposition and classification in UWB-SAR imagery. In *2018 IEEE Radar Conference, RadarConf 2018* (pp. 553–558). Piscataway: Institute of Electrical and Electronics Engineers Inc.
50. Walker, B., & Ray, L. (2019). Multi-class crevasse detection using ground penetrating radar and feature-based machine learning. In *International Geoscience and Remote Sensing Symposium (IGARSS)* (pp. 3578–3581)
51. Wang, P., Hu, Z., Zhao, Y., & Li, X. (2016). Experimental study of soil compaction effects on GPR signals. *Journal of Applied Geophysics, 126*, 128–137.

52. Warren, C., Giannopoulos, A., & Giannakis, I. (2016). gprMax: Open source software to simulate electromagnetic wave propagation for ground penetrating radar. *Computer Physics Communications, 209*, 163–170.
53. Williams, R. M., Ray, L. E., & Lever, J. H. (2012). Autonomous robotic ground penetrating radar surveys of ice sheets: Using machine learning to identify hidden crevasses. In *2012 IEEE International Conference on Imaging Systems and Techniques Proceedings* (pp. 7–12).
54. Wu, K., Rodriguez, G. A., Zajc, M., Jacquemin, E., Clément, M., De Coster, A., & Lambot, S. (2019). A new drone-borne GPR for soil moisture mapping. *Remote Sensing of Environment, 235*, 111456.
55. Yoder, R. E., Freeland, R. S., Ammons, J. T., & Leonard, L. L. (2000). Mapping agricultural fields with GPR and EMI to predict offsite movement of agrochemicals. In *Proceedings of SPIE*, vol. 4084.
56. Zhang, Y., Xiao, Z., Wu, L., Lu, X., & Wang, Y. (2017). Deep learning for subsurface penetrating super-resolution imaging. In *2017 10th UK-Europe-China Workshop on Millimetre Waves and Terahertz Technologies (UCMMT)* (pp. 1–4).
57. Zheng, J., Teng, X., Liu, J., & Qiao, X. (2019). Convolutional neural networks for water content classification and prediction with ground penetrating radar. *IEEE Access, 7*, 185385–185392.
58. Zhou, H., Feng, X., Zhang, Y., Nilot, E., Zhang, M., Dong, Z., & Qi, J. (2018). Combination of support vector machine and H-alpha decomposition for subsurface target classification of GPR. In *2018 17th International Conference on Ground Penetrating Radar, GPR 2018*. Piscataway: Institute of Electrical and Electronics Engineers Inc.

A Minimalist Approach to Yield Mapping of Standing Wheat Crop with Unmanned Aerial Vehicles

Nathaniel Narra, Antti Halla, Petri Linna, and Tarmo Lipping

Abstract Yield estimation and mapping of standing crops are often based on tedious data gathering procedures that can be daunting and not cost-effective in the absence of harvester-mounted yield mappers. A cost-effective solution with reasonable accuracy has greater potential for adoption especially if one can leverage latest machine learning tools to supplant tedious processes. This study conducts a feasibility test in using drones in a minimalist sampling strategy to estimate wheat yields over different productivity zones. The first step is to use unsupervised clustering of spatio-temporal multivariate data to delineate zones of homogeneous vegetation vigour. These zones are assumed to capture the variability in yield and aid in designing an efficient sampling strategy. The second step involves using a UAV-mounted camera to capture digital images to estimate the wheat head count and then to derive a yield estimate within the image field of view. Using physical counting of grains within a 0.5×0.5 m reference frame, the performance of the estimation procedure was observed. The results show that while the workflow is tractable and friendly in low-resource environments, the accuracy is poor at this stage. Pertinent challenges and potential improvement strategies are discussed.

Keywords Object detection · Clustering · Performance zones · FasterRCNN · Global wheat head detection · Machine learning · Yield estimation

1 Introduction

Among various activities within agriculture, crop production typically operates with an objective of maximizing yield. In this context, crop and soil management procedures are developed to optimize the yield with respect to crop inputs and interventions. It is natural then that being aware of the final yield is an important

N. Narra (✉) · A. Halla · P. Linna · T. Lipping
Tampere University, Pori, Finland
e-mail: nathaniel.narra@tuni.fi; antti.halla@tuni.fi; petri.linna@tuni.fi; tarmo.lipping@tuni.fi

T. Lipping et al. (eds.), *New Developments and Environmental Applications of Drones*,
https://doi.org/10.1007/978-3-030-77860-6_9

parameter in quantitatively monitoring and improving strategies. This parameter can be defined over a range of scales over which yield can be measured: intra-field, field, farm, regional, national and international. Each scale finds purpose in informing management decisions for practitioners, planners and policy-makers. Depending on the purpose, yield can be quantified either before harvest or after. Uses can vary from maintaining records, optimizing logistics, understanding crop response to interventions, forecasting annual crop stocks and allied statistics for food security.

There are a number of established methods for estimating the yield, of either a standing crop or harvested crop, with varying accuracy and scales. They range from manual strategies to automated measurements [28]. Among the more sophisticated methods developed for estimating yield are those that use remotely sensed spectral imagery. These involve using machine learning methods to establish correlation between the spectral properties of crops and yield [18, 19, 23]. With developments in satellite-based remote sensing, the capacity to perform analyses of cropland images has improved. In the past 2 decades, there have been a number of satellite missions that have embraced the open access data concept. This abundance of multi-year multi-spectral data from such missions has enabled large-scale development of algorithms for data analytics. Current capabilities of satellite imagery have enabled collecting and analyzing data with spatial resolutions down to 10 m. This makes it possible to estimate yields at the field scale with coarse spatial variations [7].

Capturing spatial variations of yield within a field is important from a precision agriculture perspective where optimizing crop production places equal emphasis on mitigating environmental damage, optimizing use of natural resources and maximizing yield [20, 22]. Mapping finer spatial variations requires a denser sampling of yield measurements, which can be tedious through manual methods. This is where strategies that use machine learning for an informed sampling strategy and computer vision to ascertain yield can give promising results. Any sampling strategy aims to capture the shape of the function it is sampling. In the context of mapping crop yield, sampling should be conducted at locations that capture the underlying variability in yield. The non-homogeneous distribution of factors that affect yield, such as soil type/texture, soil moisture content, influences the spatial variability of yield at harvest [1]. With advances in sensors, there are increasing spatio-temporal data sources that log soil and plant characteristics in greater detail across space and time. Advanced clustering algorithms, a branch of unsupervised machine learning, can be used to identify meaningful homogeneous clusters, also called productivity zones [4, 15]. In the absence of soil surveys and mapping of other significant factors, crop vigour observed through spectral images can be used as a proxy for calculating the underlying clusters of homogeneous crop productivity.

Sensor developments combined with positioning systems have resulted in harvesters equipped with yield mappers that measure yield during harvest at high resolutions. However, such sensor packages can be expensive. A combination of machine vision and artificial intelligence along with high-resolution spectral imagery can be used to produce high-resolution yield maps [21]. Such imagery is achievable due to developments in remote sensing platforms such as unmanned aerial vehicles (UAV) and affordable consumer grade multi-spectral cameras. UAVs provide users with control over scheduling over-flights and thus can be useful platforms for in-season crop monitoring. However, current capabilities of drones are not always cost-effective especially in terms of time and expertise required for data acquisition and its preparation for further analysis. Machine learning can be applied to extract the yield information from spectral images through learned correlations. In crops where the yield is apparent in images of crop canopy, advances in machine vision have been used in detection toward measuring yield. Thus, there have been studies where yield has been estimated using digital colour images for rice [16] and wheat [6, 32]. Thermal images of wheat canopy seem to offer better wheat head contrast, making their detection easier [5]. An alternative to handheld camera setups are UAV-mounted cameras for image data acquisition. With increasing choices of UAV-mountable high-resolution cameras (RGB, multi-spectral, thermal and even hyperspectral), conducting data acquisition over greater ranges becomes possible [8]. With developments in deep learning architectures, they have found use in detecting wheat grains [17, 30]. Among them, convolutional neural networks have been directly applied on images to detect wheat heads in images [12, 27].

While UAV-based data mapping presents advantages over satellite-based imagery—e.g., better resolution and indifference to cloud coverage—in some situations, it can be impractical. This is mostly due to the rather laborious data post-processing procedure necessary to convert multiple images into a single geo-referenced mosaic. This study aims to assemble a machine-learning-based workflow to rapidly estimate crop yield and map its spatial variability through digital images of the crop canopy. A secondary aim of the study is to identify procedures and challenges in data collection and analysis, necessary for developing a database for training and testing deep learning architectures. This non-intensive low-resource approach to the wheat head counting task is developed as a cost-effective alternative to harvester-mounted commercial yield mapper.

2 Material

The study area is a 4.19 ha field (Fig. 1) with wheat crop in 2020 season, located in the Satakunta region of Finland (lat. 60°58'43"N, long. 22°32'48"E). The field is situated in cropping zone 1 as categorized by the Natural Resources Institute Finland (Luonnonvarakeskus) [14]. The soil types present are silty moraine and sandy moraine. The final yield (per hectare) as reported by the farmer was 4400 kg/ha.

2.1 Data Sources

The remote sensing data acquired consisted of multi-spectral satellite images and drone-mounted multi-spectral data. Satellite data from Sentinel2 L2A product was

Fig. 1 Aerial orthophotograph of the field under study (boundary demarcated in dashed black lines). The points (P1, P2, P3) are locations where manual yield sampling and drone imaging were conducted. Two 0.5 × 0.5 m sampling sites were chosen at each location

acquired for 2020 growing season through the Sentinel EO Browser service.[1] Images without cloud occlusions or shadows over the field were manually selected. Six spectral bands were selected—blue, green, red, NIR, red edge and SWIR—with a bounding box covering an area of 615 × 425 m. The selected data are subsequently downloaded as a compressed folder. The UAV setup consisted of an AIRINOV 3DR Solo Quadcopter (Airinov SAS, Paris, France) mounted with a SEQUOIA multi-spectral camera (Parrot SA, Paris, France) with a built-in illumination sensor. The UAV was flown manually using the control joystick, flying at altitudes between 2 and 10 metres. An aluminium frame of 0.5 × 0.5 m was constructed to be used as a reference frame. For training wheat head detection, the global wheat head detection (GWHD) dataset was used [3]. Briefly, the dataset consists of 4700 crop canopy images with 190,000 annotations or bounding boxes with the heads at different stages of maturity.

3 Methods

The procedure involves data collection from various sources and associated tasks in processing them for analysis. A geospatial database is chosen for storing data due to possibility of spatial queries. Machine learning is applied for the critical tasks of calculating productivity zones and identifying wheat heads from images of crop canopy taken using a UAV.

[1] https://apps.sentinel-hub.com/eo-browser/.

3.1 Data Preparation

The pack of images downloaded via the EO Browser portal was loaded into QGIS geographic information system [25]. The images were clipped using the polygon shape file of the field. The pixel centres were extracted as point geo-locations, henceforth referred to as 'data points', producing 1280 such points within the field. The clipped images (rasters) and the data points were stored in a PostgreSQL database.

3.2 Clustering: Productivity Zones

Only clear satellite images (i.e., no clouds and shadows) of the field with vegetation evident in them were used for further processing, which for 2020 season was between 2 June and 18 August and constituted of 14 viable image stacks. In this chapter, two indices are used as indicators of crop condition: Normalized Difference Vegetation Index (NDVI) and Normalized Difference Moisture Index (NDMI). NDVI is a commonly used index to represent photosynthetic activity. It quantifies, on a scale of -1 to 1, the 'greenness' of the vegetation (usually >0.2) based on how plants reflect light at red and near infra-red wavelengths.

$$NDVI = \frac{NIR - RED}{NIR + RED}.$$

NDMI is an index that represents water content in vegetation. The values are between -1 and 1. Values between -0.2 and 0.4 typically represent water stress, and high-positive values represent canopy without water stress (>0.4).

$$NDMI = \frac{NIR - SWIR}{NIR + SWIR}.$$

A z-score normalization was applied to each index raster for a particular date. This ensures 0 mean and a standard deviation of 1 for data on each date and, instead of indices depending on the crop phenological stage, ensures a more reliable comparison temporally [22].

$$zscore_n(t) = \frac{index_n(t) - \mu_n(t)}{\sigma_n(t)}.$$

The calculated indices were stored in a PostgreSQL database as rasters and the attribute values at each data point queried. A feature vector was assembled for each data point and its attributes.

$$f_n = \{lat, lon, ndvi_n(t_1), ndmi_n(t_1), \ldots, ndvi_n(t_i), ndmi_n(t_i)\},$$

where lat, lon are the latitude and longitude of the data point n; $ndvi_n/ndmi_n(t)$ is the index value of data point n at time t ($i = 14$).

The feature vectors were then used to perform unsupervised clustering using the IBM Watson Studio Desktop application (IBM, New York, USA). Clustering was performed through the *autocluster* node that estimates and compares multiple clustering models (Two-Step, K-Means and Kohonen) by iterating through multiple combinations of options. Models were compared and ranked using the silhouette metric.

3.3 Wheat Grain Data

To gather wheat head and grain statistics, each cluster was sampled at 2 closely placed locations. At each location, the procedure followed was: the metal reference frame was placed at an appropriate location. After the site was imaged by the drone, the wheat heads within the frame were harvested and stored in marked plastic bags. The heads were subsequently left in the open at room temperature for drying. Once dried, each head was manually de-shelled and the following measures were noted: head width, head length and grain count. These detailed counts were noted for at least 100 heads for each zone. The total head count and grain weight for each location were noted as well.

3.4 Drone Imaging

To keep drone operation workload to a minimum, no pre-flight procedures such as flight path planning were performed. After placing the reference frame at a chosen location, the UAV is manually made to hover at various altitudes between 2 and 10 m with the camera taking consecutive images at regular time intervals (2 s). The images were stored on the on-board memory card within the camera housing. After the flights, the images were transferred to a workstation where the appropriate images were manually chosen. Images that were blurred or where the reference frames were too far from the centre of the image were discarded. Multiple images for each location were chosen such that they were of different perspectives and altitudes (and consequently slightly different spatial resolutions).

3.5 *Wheat Head Detection*

As the fundamental task is that of object detection—object being the wheat head—the straightforward choice was using the FasterRCNN method [26]. Briefly, the network works by first extracting features from an image. Second, the Region Proposal Network within the network generates regions that potentially contain an object. A major performance enhancement from this network comes due to the use of anchor boxes to detect images at various scales and aspect ratios. Another performance enhancer is the use of the same feature map detecting regions and classification. In this chapter, the basic PyTorch implementation includes the ResNet50 as a backbone. The code is a development of a public code[2] released on the Kaggle site for the global wheat detection challenge.[3]

4 Results

The results of the performance comparison of the clustering algorithms (Two-Step, K-means, Kohonen) showed that the standard K-means (silhouette = 0.507) and Two-Step (silhouette = 0.512) algorithm performed similarly (Fig. 2). There were 4 clusters produced with the Two-Step clustering and 5 clusters with K-means clustering.

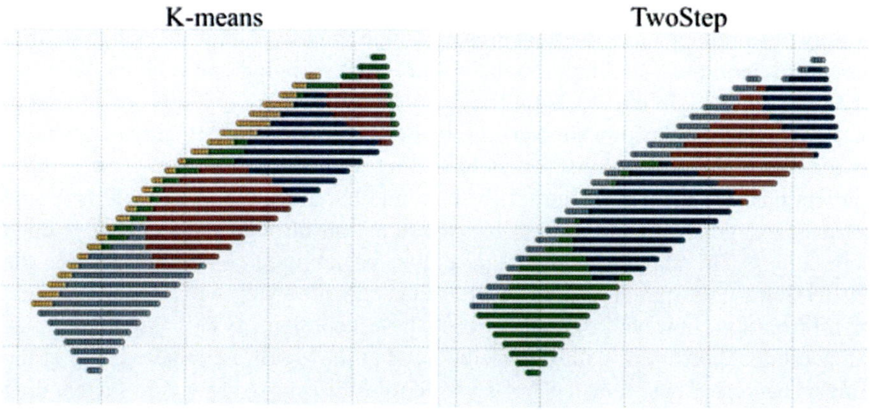

Fig. 2 Results of clustering—*left*: K-means clustering with an optimal 5 clusters; *right*: Two-Step clustering produced with 4 clusters. It is worth noting that the top 3 largest clusters are similar in extent

[2]https://www.kaggle.com/pestipeti/pytorch-starter-fasterrcnn-inference.

[3]https://www.kaggle.com/c/global-wheat-detection.

Grain sampling locations were chosen roughly in the central regions of these zones. 2 locations were chosen in each zone for a little more robustness in the cumulative statistics. Table 1 lists the various parameters measured at the chosen sample locations. Plots illustrating the distribution of grains per head versus head lengths and widths, along with their marginal distributions, are shown in Fig. 3.

Five images were chosen from the image stack for each of the six sampling locations. Bulk counts of wheat heads when compared to manual measurements showed an underestimation of the counts (Fig. 4). The FasterRCNN architecture used has not been tuned extensively and only been trained on 4700 images with plants in differing phenological stages.

In terms of computation, none of the modules is resource-intensive once the detection architecture is trained. The most labour-intensive part of the training work flow remains the manual sampling of the wheat crop, which is an essential task for training the network.

5 Discussion

The importance of estimating, measuring or forecasting crop yield is made apparent due to its immediate economic significance. It can be further useful as a tentative indicator of the state of the ecosystem where interactions within its components—such as soil, soil biota, weather/climate and water—have a bearing on crop growth and subsequently the final yield. This yield can be measured at various scales depending on the resolution at which data are collected. Thus, depending on the density of sampling and scale of aggregation, one can ascertain yield using satellite-based estimation of standing crop, harvester-based mapping and regional statistics of harvested yield. In the precision farming paradigm, where judicious optimization of yield is desired through site-specific management practices, it can be beneficial to map the spatial variability of yield at high resolution. Typically, there are a few options: (a) manual counting, (b) combine harvester-mounted yield mapping sensor package and (c) algorithm-based estimation through modelled or learned correlations. The manual methods are either too laborious or do not provide the desired intra-field spatial resolution. Harvester-mounted yield sensors can map at a high resolution; however, equipment costs have kept them from being ubiquitous. An additional constraint with this method is that the yields are mapped only at the time of harvest. Thus, they are not a viable tool for estimating yield of a standing crop. Algorithmic approaches can address these constraints by achieving reasonable accuracy and offer a more accessible solution that can be used for crop monitoring. However, the scope and capability of these algorithms can vary widely. They can range from simple linear regression models to sophisticated multivariate deep learning (non-linear) or probabilistic models that can forecast yield. The advantage to the algorithmic approach is that a significant part of the labour-intensive process is executed during the training stages. With developments in sensors and computer vision methods, many of these tedious tasks can be automated, such as using object

Table 1 Wheat head measures observed at 6 sampling locations in the 3 chosen zones. A subset of the total was individually measured for grain counts and weight. Yield (in kg/ha) per head is measured by dividing the yield potential of each sample location by the number of detected heads

Location		Head counts and yield				Extrap.	Grain measures		
		Head #	Weight (gms)	Yield (kg/ha)		Field avg. (kg/ha)	Head #	Grain #	Weight (gms)
P1	1	204	45	1800	1780	5573	–	–	–
	2	245	44	1760			–	–	–
P2	1	243	248	9920	8500		70	1467	75
	2	168	177	7080			50	1200	57
P3	1	165	146	5840	6440		70	1505	70
	2	195	176	7040			75	1571	75

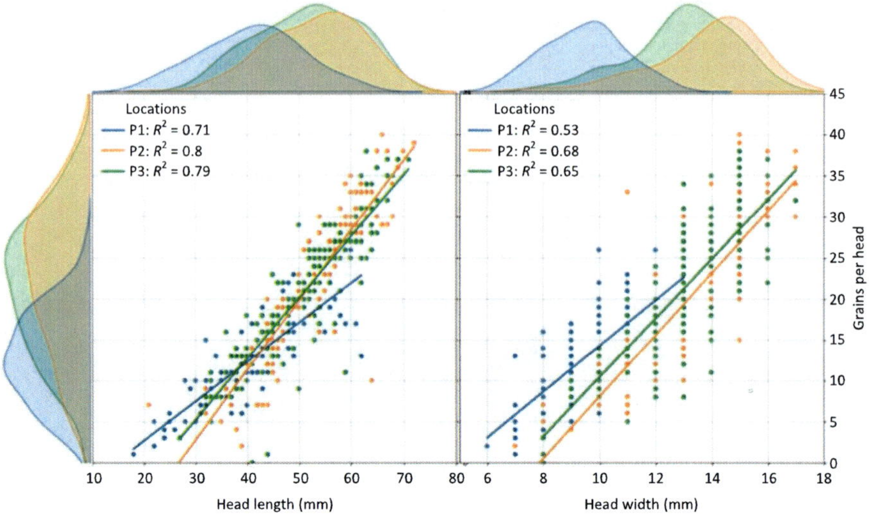

Fig. 3 Descriptive scatter plots with regression lines and marginal plots for—*left*: head length versus grain count per head for each zone (P1, P2, P3). The counts observed for the 2 sampling locations in each zone are combined; *right*: similar plot of head width versus grain count per head

detection instead of manual counting, while UAVs enable quick access to sampling locations.

In this chapter, an attempt has been made to ascertain the tractability of a low-labour approach to estimating yield of standing wheat crop in a field. The UAV-mounted camera-based workflow described in Fig. 5 is designed to supplant resource heavy procedures—in terms of labour, time or other resources—with tools developed in the machine learning and computer vision domain. The overall approach is to select sampling locations that optimally capture the yield heterogeneity within the field, use computer vision tools to detect wheat heads in a size-standardized crop patch and finally extrapolate the estimated patch yield over the entire field.

Each of these steps contributes toward resource efficiency through assumptions made about the relationship between direct measures and their proxies. The major assumptions are: (1) grain weight can be correlated to wheat head counts and morphometrics [11]; (2) clustering of vegetation indices captures yield heterogeneity. Technological and procedural limitations impose certain constraints on the workflow: (1) drone flights are limited by regulatory and weather-dependent flight constraints; (2) camera capabilities determine imaging parameters and perspective distortion; (3) identifying performance zones is dependent on the availability of cloud-free satellite time series images (4) image-based detection and morphometry is heavily dependent on training data volume. While the results of this study show that the yields estimated are extremely low, it should, however, be noted that the relative differences in yield among the three locations are similar. In other words,

P1

Site	Image sample					Mean	Measured
	N_1	N_2	N_3	N_4	N_5	$\overline{N}$	N_{true}
1	36	43	49	29	44	40	204
2	43	55	63	42	57	52	245

Measured yield: 1780 *kg/ha*
Estimated yield: 369 *kg/ha*

P2

Site	Image sample					Mean	Measured
	N_1	N_2	N_3	N_4	N_5	$\overline{N}$	N_{true}
1	41	55	53	56	49	50	243
2	40	59	57	52	55	53	168

Measured yield: 8500 *kg/ha*
Estimated yield: 2145 *kg/ha*

P3

Site	Image sample					Mean	Measured
	N_1	N_2	N_3	N_4	N_5	$\overline{N}$	N_{true}
1	42	50	54	53	45	49	165
2	44	43	39	36	51	43	195

Measured yield: 6440 *kg/ha*
Estimated yield: 1645 *kg/ha*

Fig. 4 Results of running the FasterRCNN architecture over data with inferred wheat heads demarcated in red boxes. 3 sample images are shown. 5 chosen images at each of the 6 sampling locations were tested. The total number of heads detected for each of the images is displayed in the table below along with the manually observed true counts (red bold font). The corresponding yield for each zone (P1, P2, P3) is calculated based on the average number of heads detected per zone and the corresponding yield/head (from Table 1). Images of samples of grains gleaned from the respective locations are shown to illustrate the difference in grain sizes

the estimation is low by similar margins. In summary, the results help identify 4 modules that are essential for improvements and critical for performance.

- **Clustering** methods provide a coherent way of uncovering productivity zones that can help increase the confidence in estimations that are extrapolated from sample observations. These zones can help in making decisions on the number of samples to take and their locations, which are necessary to capture the variation in the field. In this chapter, two indices were used as a basis for partitioning the field into clusters—NDVI, NDMI. The NDVI exhibits the vegetative vigour that can give information about variations in early crop development. With NDMI, it is expected that variations in plant water stress states and differential rates of senescence will expose the distribution of underlying factors. One factor to account for are edge effects at field boundaries due to resolution differences in the different sentinel bands. Methods as proposed by Yamada and Rogerson [31] should be incorporated to get the most relevant clusters. The clustering algorithms used in this study, the simple K-means-based process, have been

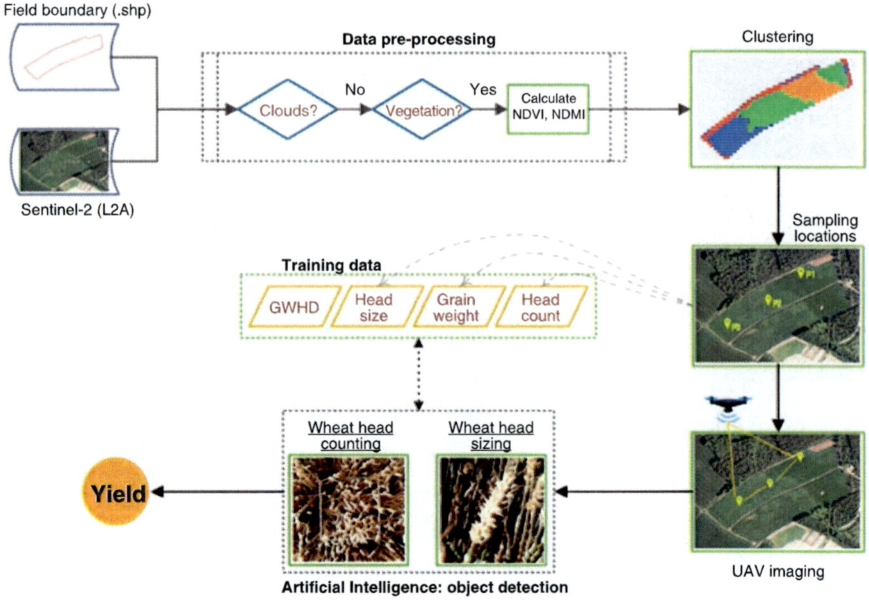

Fig. 5 Schematic of the conceptual workflow including data sources

shown to be effective. However by using functional clustering methods, more rational discrimination of clusters in time series data can be achieved [22].

- **Training data—images**: In spite of the availability of an excellent dataset (GWHD: 4700 images), more training image samples can certainly help with training algorithms. Additionally, placing a 0.5 m reference frame within the image field of view would not be practical during training data acquisition. Such a frame can be inferred from images. By detecting the crop heights under the camera using a LiDAR [29], the optical geometry and distortion of the specific camera can be used to calculate the distances in the images. An alternative approach can be using lasers to project frames within the field of view of the images [13]. Though, how the uneven nature of the crop canopy affects the laser lines remains to be studied.
- **Head detection**: There is potential for significant improvement in the object detection task, where the detection rate has been very low ($20 - 30\%$) leading to a corresponding underestimation of yield. Certainly some detection error is due to the fact that in the images the wheat heads and vegetation at the canopy level tend to occlude the heads below. However, better detection can be expected if the FasterRCNN architecture can be optimized and trained with more data. Additional deep learning architectures, such as the recent Yolov4 architecture [2], can also be explored for improved performance [10].

- **Head morphometry**: The results indicate that head counts alone are not good predictors of yield. However, when considered along with their lengths and widths, they can potentially be correlated to yield. Manually collecting a distribution of these features is obviously not an efficient solution. Image-based morphometry [9] can be implemented to gather the essential length and width features. Performing precise measurements with a UAV from an overhead perspective may pose significant challenges, due to inability to image heads from the side. However, by using the downdraft of the UAV rotors to change the relative position of the head to the camera could be a possible approach. This would require concerted effort in gathering training data to understand the confidence with which morphology can be estimated using such a technique.

In the future, a data collection campaign will be designed based on the improvement strategies discussed above, in part to address a major limitation of the GWHD database. In the context of this study, the database does not provide true counts in the crop patch within the image view; instead, it only provides locations of image detectable heads (un-occluded). Nor does it correlate these images to yield. The results of this study seem to indicate that using detected counts of wheat heads directly in calculating yield is highly dependent on detection algorithm performance. Unless image acquisition protocols can capture all existing heads present in field of view by overcoming physical occlusions, it may be prudent to follow a more probabilistic approach. Using the Bayesian framework to develop prior distributions of head metrics and yield can possibly help overcome certain limitations of the detection process. This justifies more data along the lines of GWHD database [3], but with additional physical measures of wheat heads and grains and yields. Further efforts will be made to gather studies that report on grain distribution and their yield potential [24] and those that ideally extend the scope of investigation to wheat head lengths and widths. Minor improvements can also be made to smooth the transfer of data between various components of the workflow, such as automating the data transfer between the camera and the workstation.

In conclusion, this chapter outlines the essentials of an important tool for research data collection campaigns. It should be noted that the foreseeable utility of this is not to replace current methods of estimation or measurement. Traditional calibrated methods are important to gather 'ground truth' values that are very low on procedure complexity though high on labour cost. Satellite-based and other remote-sensing-based measures have higher resolutions but also involve a more involved data workflow with processes that can be complex. Yield mappers provide reliable measures with high resolution; however, they are a significant investment as well as limited in scope to harvested yield. With the approach suggested here, there are two main uses: practitioners without yield mappers who would like to estimate yield of standing crop or at the time of harvesting to gather a reasonable distribution of yield. The other use case, and the primary motivation for this study, is that when conducting research on test fields where access to measuring methods is varied between different participating farmers, it can be a useful tool for researchers to be able to assess and monitor yields for record keeping and monitoring the effects of interventions across time.

References

1. Ansari, M. Y., Ahmad, A., Khan, S. S., Bhushan, G., & Mainuddin (2020). Spatiotemporal clustering: A review. *Artificial Intelligence Review, 53*, 2381–2423.
2. Bochkovskiy, A., Wang, C. Y., & Liao, H. Y. M. (2020). YOLOv4: Optimal Speed and Accuracy of Object Detection. arXiv:2004.10934.
3. David, E., Madec, S., Sadeghi-Tehran, P., Aasen, H., Zheng, B., Liu, S., Kirchgessner, N., Ishikawa, G., Nagasawa, K., Badhon, M. A., Pozniak, C., de Solan, B., Hund, A., Chapman, S. C., Baret, F., Stavness, I., & Guo, W. (2020). Global wheat head detection (GWHD) dataset: A large and diverse dataset of high-resolution rgb-labelled images to develop and benchmark wheat head detection methods. *Plant Phenomics, 2020*, 3521852 .
4. Ertl, B., Meyer, J., Streit, A., & Schneider, M. (2019). Application of mixtures of gaussians for tracking clusters in spatio-temporal data. In *Proceedings of the 11th International Joint Conference on Knowledge Discovery, Knowledge Engineering and Knowledge Management* (vol. 1, pp. 45–54). SCITEPRESS - Science and Technology Publications.
5. Fernandez-Gallego, J., Buchaillot, M., Gutiérrez, N. A., Nieto-Taladriz, M., Araus, J., & Kefauver, S. (2019). Automatic wheat ear counting using thermal imagery. *Remote Sensing, 11*, 751.
6. Fernandez-Gallego, J. A., Kefauver, S. C., Gutiérrez, N. A., Nieto-Taladriz, M. T., & Araus, J. L. (2018). Wheat ear counting in-field conditions: High throughput and low-cost approach using RGB images. *Plant Methods, 14*(22).
7. Fieuzal, R., Bustillo, V., Collado, D., & Dedieu, G. (2020). Combined use of multi-temporal landsat-8 and sentinel-2 images for wheat yield estimates at the intra-plot spatial scale. *Agronomy, 10*, 327.
8. Fu, Z., Jiang, J., Gao, Y., Krienke, B., Wang, M., Zhong, K., Cao, Q., Tian, Y., Zhu, Y., Cao, W., & Liu, X. (2020). Wheat growth monitoring and yield estimation based on multi-rotor unmanned aerial vehicle. *Remote Sensing, 12*, 508.
9. Genaev, M. A., Komyshev, E. G., Smirnov, N. V., Kruchinina, Y. V., Goncharov, N. P., & Afonnikov, D. A. (2019). Morphometry of the wheat spike by analyzing 2D images. *Agronomy, 9*(7), 390.
10. Gong, B., Ergu, D., Cai, Y., & Ma, B. (2021). Real-time detection for wheat head applying deep neural network. *Sensors, 21*(1).
11. Guo, Z., Zhao, Y., Röder, M. S., Reif, J. C., Ganal, M. W., Chen, D., & Schnurbusch, T. (2018). Manipulation and prediction of spike morphology traits for the improvement of grain yield in wheat. *Scientific Reports, 8*, 1–10.
12. Hasan, M. M., Chopin, J. P., Laga, H., & Miklavcic, S. J. (2018). Detection and analysis of wheat spikes using convolutional neural networks. *Plant Methods, 14*(1), 100.
13. Isop, W. A., Pestana, J., Ermacora, G., Fraundorfer, F., & Schmalstieg, D. (2016). Micro aerial projector - stabilizing projected images of an airborne robotics projection platform. In *2016 IEEE/RSJ International Conference on Intelligent Robots and Systems (IROS)* (pp. 5618–5625).
14. Laine, A., Hügnäsbacka, M., Niskanen, M., Ohralahti, K., Jauhiainen, L., Kaseva, J., & Nikander, H. (2017). Results of the official variety trials 2009–2016. Retrieved from http://urn.fi/URN:ISBN:978-952-326-346-8
15. Lamb, D. S., Downs, J., & Reader, S. (2020). Space-time hierarchical clustering for identifying clusters in spatiotemporal point data. *ISPRS International Journal of Geo-Information, 9*, 85.

16. Liu, K., Li, Y., Han, T., Yu, X., Ye, H., Hu, H., & Hu, Z. (2019). Evaluation of grain yield based on digital images of rice canopy. *Plant Methods, 15*, 28.
17. Madec, S., Jin, X., Lu, H., Solan, B. D., Liu, S., Duyme, F., Heritier, E., & Baret, F. (2019). Ear density estimation from high resolution RGB imagery using deep learning technique. *Agricultural and Forest Meteorology, 264*, 225–234.
18. Nevavuori, P., Narra, N., Linna, P., & Lipping, T. (2020). Crop yield prediction using multitemporal UAV data and spatio-temporal deep learning models. *Remote Sensing, 12*(23), 4000.
19. Nevavuori, P., Narra Girish, N., & Lipping, T. (2019). Crop yield prediction with deep convolutional neural networks. *Computers and Electronics in Agriculture, 163*, 104859.
20. Ohana-Levi, N., Bahat, I., Peeters, A., Shtein, A., Netzer, Y., Cohen, Y., & Ben-Gal, A. (2019). A weighted multivariate spatial clustering model to determine irrigation management zones. *Computers and Electronics in Agriculture, 162*, 719–731.
21. Pantazi, X. E., Moshou, D., Alexandridis, T., Whetton, R. L., & Mouazen, A. M. (2016). Wheat yield prediction using machine learning and advanced sensing techniques. *Computers and Electronics in Agriculture, 121*, 57–65.
22. Pascucci, S., Carfora, M., Palombo, A., Pignatti, S., Casa, R., Pepe, M., & Castaldi, F. (2018). A comparison between standard and functional clustering methodologies: Application to agricultural fields for yield pattern assessment. *Remote Sensing, 10*, 585.
23. Patrício, D. I., & Rieder, R. (2018). Computer vision and artificial intelligence in precision agriculture for grain crops: A systematic review. *Computers and Electronics in Agriculture, 153*, 69–81.
24. Philipp, N., Weichert, H., Bohra, U., Weschke, W., Schulthess, A. W., & Weber, H. (2018). Grain number and grain yield distribution along the spike remain stable despite breeding for high yield in winter wheat. *PLoS ONE, 13*(10), e0205452.
25. QGIS Development Team. (2021). *QGIS Geographic Information System*. QGIS Association.
26. Ren, S., He, K., Girshick, R., & Sun, J. (2015). Faster R-CNN: Towards real-time object detection with region proposal networks. arXiv:1506.01497.
27. Sadeghi-Tehran, P., Virlet, N., Ampe, E. M., Reyns, P., & Hawkesford, M. J. (2019). Deepcount: In-field automatic quantification of wheat spikes using simple linear iterative clustering and deep convolutional neural networks. *Frontiers in Plant Science, 10*, 1176.
28. Sud, U., Ahmad, T., Gupta, V., Chandra, H., Sahoo, P. M., Aditya, K., Singh, M., & Biswas, A. (2017). Methodology for estimation of crop area and crop yield under mixed and continuous cropping. Retrieved from http://www.fao.org/3/ca6514en/ca6514en.pdf
29. ten Harkel, J., Bartholomeus, H., & Kooistra, L. (2020). Biomass and crop height estimation of different crops using uav-based lidar. *Remote Sensing, 12*(1), 17.
30. Wu, W., le Yang, T., Li, R., Chen, C., Liu, T., Zhou, K., ming Sun, C., yan Li, C., kai Zhu, X., & shan Guo, W. (2020). Detection and enumeration of wheat grains based on a deep learning method under various scenarios and scales. *Journal of Integrative Agriculture, 19*, 1998–2008.
31. Yamada, I., & Rogerson, P. A. (2003). An empirical comparison of edge effect correction methods applied to k-function analysis. *Geographical Analysis, 35*(2), 97–109.
32. Zhou, C., Liang, D., Yang, X., Yang, H., Yue, J., & Yang, G. (2018). Wheat ears counting in field conditions based on multi-feature optimization and twsvm. *Frontiers in Plant Science, 9*, 1024.

Assessment of Crop Yield Prediction Capabilities of CNN Using Multisource Data

Petteri Nevavuori, Nathaniel Narra, Petri Linna, and Tarmo Lipping

Abstract The growing abundance of digitally available spatial, geological, and climatological data opens up new opportunities for agricultural data-based input–output modeling. In our study, we took a convolutional neural network model previously developed on Unmanned Aerial Vehicle (UAV) image data only and set out to see whether additional inputs from multiple sources would improve the performance of the model. Using the model developed in a preceding study, we fed field-specific data from the following sources: near-infrared data from UAV overflights, Sentinel-2 multispectral data, weather data from locally installed Vantage Pro weather stations, topographical maps from National Land Survey of Finland, soil samplings, and soil conductivity data gathered with a Veris MSP3 soil conductivity probe. Either directly added or encoded as additional layers to the input data, we concluded that additional data helps the spatial point-in-time model learn better features, producing better fit models in the task of yield prediction. With data of four fields, the most significant performance improvements came from using all input data sources. We point out, however, that combining data of various spatial or temporal resolution (i.e., weather data, soil data, and weekly acquired images, for example) might cause data leakage between the training and testing data sets when training the CNNs and, therefore, the improvement rate of adding additional data layers should be interpreted with caution.

Keywords Crop yield prediction · CNN · Multisource input · Remote sensing · Intra-field

P. Nevavuori (✉)
Mtech Digital Solutions Oy, Vantaa, Finland
e-mail: petteri.nevavuori@mtech.fi

N. Narra · P. Linna · T. Lipping
Tampere University, Tampere, Finland
e-mail: nathaniel.narra@tuni.fi; petri.linna@tuni.fi; tarmo.lipping@tuni.fi

T. Lipping et al. (eds.), *New Developments and Environmental Applications of Drones*,
https://doi.org/10.1007/978-3-030-77860-6_10

1 Introduction

The application of novel and performant deep learning techniques has seen an increasing trend in the last few years in the domain of Smart Farming and Precision Agriculture [16]. Multiple factors are at play: the abundance of open access satellite system spatial data, availability of commercial unmanned aerial vehicles (UAVs) mountable with external sensors, developments in the soil sensor and camera sensor technologies, and the constant need to optimize the production of farms.

Convolutional neural networks (CNNs), being a subset of deep learning, have been utilized in recent studies on crop yield prediction [16]. The spatial model architecture has been used in predicting cotton yield from RGB data taken at close proximity [14], cereal crop yield prediction from mid-altitude UAV RGB data [9], rice grain yield estimation [18], and crop yield prediction using multisource inputs on patch scale [2]. In [9], we compared intra-field crop yield estimation performance with NDVI and RGB data from the earlier and later part of the growing season with a variety of CNN configurations. The focus of that study was to assess the generalization capability of a yield prediction model with UAV RGB data.

1.1 Objectives

In this study, we examine the effect of additional field-related spatial or spatial-like data on the intra-field crop yield prediction capabilities using data gathered from the earlier half of the growing season of 2018 (weeks 21–26). The objective of this study is to assess crop yield prediction capabilities with the best CNN model composition from [9] by varying the input data configuration. The focus of this study is to see whether additional data, such as weather data, soil and ground information, and open-access Sentinel-S2 data would improve the point-in-time prediction performance compared to just using UAV-based RGB data. To limit the scope of the study, architectural and hyperparameter tuning of the CNN model is not addressed here to better isolate performance changes to data, and the tuned out architectural and optimizer-related hyperparameters were thus taken from [9].

2 Material and Methods

2.1 Data Acquisition

For this study, four crop fields were selected for data acquisition in the vicinity of Pori, Finland (61°29′6.5″N, 21°47′50.7″E) for the growing season of 2018. The field information is provided in Table 1. Following the conclusions of [9], only data

Table 1 The fields selected for the study in the proximity of Pori, Finland. The thermal time is calculated as the cumulative sum of temperature between the sowing and harvest dates. Mean yield has been calculated from processed yield sensor data for each field

Field no.	Size (ha)	Mean yield (kg/ha)	Crop (*Variety*)	Thermal time (°C)	Sowing date
1	7.59	5157.6	Wheat (*Mistral*)	1316.8	14 May
2	11.77	5534.3	Barley (*Zebra*)	1179.9	12 May
3	7.88	4166.9	Barley (*RGT Planet*)	1127.6	16 May
4	7.24	6166.0	Oats (*Ringsaker*)	1216.4	18 May

Table 2 General information of data sources and their original formats

Source	Type	Resolution/step	Multitemporal
UAV	Raster	0.3125 m/px	Yes
Sentinel-S2	Raster	[10, 20, 60] m/px	Yes
Soil samples	Vector	50 m	No
Veris MSP3	Vector	20 m	No
Topography	Vector	2 m	No
Weather	Tabular	–	Yes
Yield	Vector	Varying	No

from the earlier half of the growing season was considered for UAV and Sentinel-S2 data.

The multisource input data for the fields consists of UAV-based RGB images, location data, multispectral Sentinel-2 [3] satellite data, sparsely collected and analyzed soil samplings, machine-collected soil information, topography information, and local weather station data. General information about the original data sources is given in Table 2. Some of the data were collected during the growing season of 2018 either manually or automatically, while other data were acquired within 1-year time difference from the aforementioned season. A total of 39 layers constitute the input data sets, while a single layer, the crop yield, is used as the ground truth. These data are described next and the data layers are numbered for further reference.

2.1.1 UAV

It has already been demonstrated that UAV-based RGB data from the first half of the growing season works better than the data from the second half of the growing season and better than NIR only in crop yield prediction [9]. The UAV data of this study has also been used in [8]. The images were taken at an average height of 150 m with a minimum of three ground control points for geometric calibration. Color correction was performed preflight and illumination sensors were used for radiometric calibration. We selected UAV-based RGB data acquired for the first weeks after sowing (weeks 21–26 of 2018). Thus, every imaged field has five distinct UAV RGB rasters in the collected data set. The data were acquired with overflights using a SEQUIOA (Parrot Drone SAS, Paris, France) multispectral camera mounted on a Airinov Solo 3DR (Parrot Drone SAS, Paris,

France) UAV. Field-wise orthomosaics were constructed with Pix4D (Pix4D S.A., Prilly, Switzerland) software. UAV data contain the following layers:

1. Red
2. Green
3. Blue.

2.1.2 Sentinel-S2

The Sentinel-S2 satellite data for the fields were acquired from the Copernicus Open Access Hub (European Space Agency, Paris, France). The data were date-matched to UAV images during acquisition, prioritizing images where the algorithmically determined cloud probability was lowest. Thus, five Sentinel-S2 rasters with temporal spacing similar to the UAV data were selected for the data set. With the abbreviated names of product layers in brackets, the Level-2A Sentinel-S2 consists of the following layers:

4. Wavelength 0.443 μm (B01)
5. Wavelength 0.490 μm (B02)
6. Wavelength 0.560 μm (B03)
7. Wavelength 0.665 μm (B04)
8. Wavelength 0.705 μm (B05)
9. Wavelength 0.740 μm (B06)
10. Wavelength 0.783 μm (B07)
11. Wavelength 0.842 μm (B08)
12. Wavelength 0.865 μm (B8A)
13. Wavelength 0.945 μm (B09)
14. Wavelength 1.610 μm (B11)
15. Wavelength 2.190 μm (B12)
16. Aerosol optical thickness at 550 nm (AOT)
17. Scene classification layer (SCL)
18. Water vapor map (WVP)
19. Cloud probability (CLDPRB)
20. True color, red (TCIR)
21. True color, green (TCIG)
22. True color, blue (TCIB).

2.1.3 Soil Samples

Soil samples were manually collected from the fields by ProAgria, an agronomic counseling institution, and sent to a Eurofins (Eurofins Viljavuuspalvelu, Mikkeli, Finland) laboratory for further analysis. Soil samples were collected with 50 m steps so that a single sample represented an area of 50 × 50 m. The samples were collected manually once during November 2018. Being point vectors, the data were rasterized with the `gdal_warp` program of the GDAL utility [17]. Soil sample data contain the following layers:

23. Calcium
24. Copper
25. Potassium
26. Magnesium
27. Manganese
28. Phosphorus
29. Sulfur
30. Zink.

2.1.4 Veris MSP3

To get a finer map of soil characteristics, a MSP3 soil scanner (Veris Technologies, Salina, Kansas, USA) was used to map the fields at depths of 0–30 cm and 30–90 cm. The measurements were performed during April and May of 2019. The MSP3 measures the soil's electrical conductivity (EC), which is an indicator of soil compactness, wetness, and soil type proportions. Additionally, the instrument measures the pH of the soil. Being irregularly spaced point data initially, data had to be rasterized from point vectors. The rasterization was done with the `gdal_warp` program of the GDAL utility [17]. Each field was measured once. Veris MSP3 data contain the following layers:

31. Shallow EC
32. Deeper EC
33. Ratio, (EC SH / EC DP)
34. Infra-red reflectance
35. Red reflectance
36. Soil pH.

2.1.5 Topography

The National Land Survey of Finland conducts light detection and ranging (LiDAR)-based elevation mappings on a regular basis in Finland. This data is openly available for anyone to download [10] and contains laser-scanned point cloud data with approximately one point per $2\,m^2$ [7]. The LiDAR data set was acquired for each of the four fields. The LiDAR data were converted from point cloud data to spatial rasters using the ArcGIS (Esri, Redlands, California, USA) software. During the conversion, the data were interpolated to match UAV data in terms of resolution. The topography data contain only the following layer:

37. Elevation information.

2.1.6 Weather Data

Weather data were collected with two separately located Vantage Pro2 (Davis Instruments, Hayward, California, USA) weather stations. As the fields constitute two distinct clusters, a weather station was placed in the immediate vicinity of each field cluster. While the stations log multiple variables with a time resolution of just minutes, we utilized accumulated daily statistics and matched data to UAV acquisition dates. Thus, five weather data maps were constructed for each field spacing matching the dates of the UAV data. The weather data contain the following layers:

38. Cumulative temperature sum
39. Cumulative rain sum.

2.1.7 Yield Data

As the task of regression is that of supervised prediction, the training of the CNN model requires information about the ground truth, the target values. These were acquired during the harvest of 2018 via yield mapping sensor devices attached to the harvesters, either with a CFX 750 (Trimble Navigation, Sunnyvale, California, USA) or with a Greenstar 1 (John Deere, Molinde, Illinois, USA). CFX 750 utilizes optical sensors to measure yield throughput and moisture. Greenstar 1 utilizes a kinetic mass flow sensor to measure yield throughput and a separate moisture sensor. The yield maps generated by the mapping equipment were initially in the form of vector point clouds. The irregularly spaced points were filtered prior rasterization to contain only points where the yield was between 1500 and 15,000 kg/ha and the harvester speed between 2 and 7 km/h, following the yield preprocessing methodology of [9]. Rasterization was then done by interpolating the yield data to form a raster image.

2.2 *Data Preprocessing*

2.2.1 Interpolation

The first step after the acquisition of data was to harmonize the spatial resolution across multiple different sources. The UAV data were initially downsampled to 0.3125 m/px or 32 pixels per 10 m. This is to match the method of data processing in [9]. Main reasons are to limit the inputs to reasonable size and to have the input dimensions conform to a power of 2 for GPU-based computations. The coarser data, namely Sentinel-S2, soil samples, Veris MSP3, elevation, and yield data, required upsampling via interpolation to match this resolution. The interpolation was done using the GDAL utility's `gdal_grid`

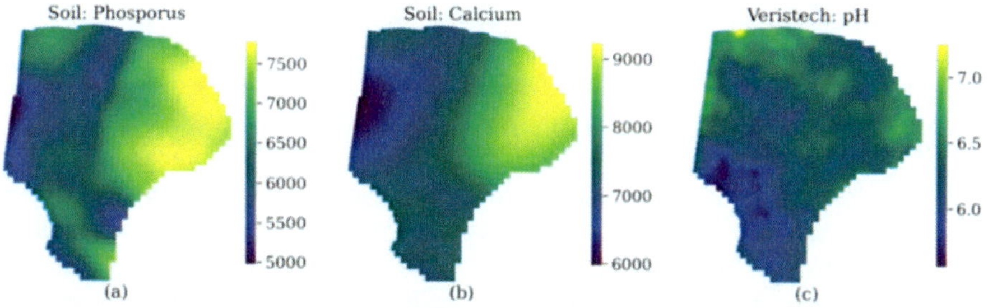

Fig. 1 Examples of input data interpolations on field scale. (**a**) the interpolated phosphorus map, (**b**) the interpolated calcium content in the field, and (**c**) the pH map as measured by the Veris MSP3 soil mapper

program with `invdist:power=3:smoothing=20` interpolation algorithm. As with the input data, also the target crop yield data were interpolated to UAV matching resolution. Example results of interpolation are depicted in Fig. 1.

2.2.2 Input Feature Normalization

After interpolation, the next step was to normalize the data. While absolute values could also be directly used, scaling the input values close to the magnitude of the model's parameters (i.e., weights) helps the model converge faster. Input layers were normalized using a function

$$d^{NORM} = (d - \mu_d)/(max(d) - min(d)), \quad d \in D, \tag{1}$$

where d is a layer in the set of all layers D in the data set and d^{NORM} is the normalized layer. However, the target crop yield values were not scaled, akin to [9].

2.2.3 Frame Separation

The CNNs require input data to have fixed dimensions. Low number of fields and the irregularities of field shapes led us to extract smaller, fixed dimension frames from the field data. Following [9], we extracted overlapping 40 × 40 m (128 × 128 px) frames with 10 m horizontal and vertical steps. Prior extraction, all input and target data from various input sources were aligned in terms of geolocation and resolution to ensure frame extraction from matching areas. Frames containing half or more valid pixels were included in data, while those having less than half were discarded. This resulted in a total of 16,375 input target frames.

2.2.4 Data Sets

Extracted samples were then divided into training, validation, and test sets. Training and validation sets were utilized during the training, while the test set was set aside as the out-of-sample performance evaluation data set. As the number of unique fields was low, we wanted to maximize the sample variability the model sees during training. We first attempted to train the models with data separated on a per-field basis with two fields for training, one for validation and one for testing. This led to the model overfitting to the training data and poor generalization performance due to low training data set variability. Similarly, a low performance was achieved with splitting fields to separate training, validation, and test sections. Thus, we then decided to divide the data temporally into distinct training, validation, and test sets according to the UAV image acquisition week. The samples were then shuffled to eliminate spatial autocorrelation in subsequent samples due to overlapping frame extraction. Used weeks, sample counts, and sample count proportions for separated sets are given in Table 3.

2.3 Model Architecture

Convolutional neural networks (CNNs) are a subset of spatial model architectures within the broader context of deep learning. CNNs excel in tasks, where the inputs fed to the model are either images or image-like data, i.e., spatial data [5, 13]. While the inner workings of the CNNs have already been well documented [9], we quickly review the operating principles of a CNN. The architecture operates with layers, like many of the deep learning architectures. Each layer is a combination of a convolution operation, which is often followed by a pooling operation. At the heart of the model are the trainable filters of the convolution operation, i.e., the kernels, which produce feature maps for further use.

In our study, we implement and use the best performing CNN architecture of [9]. The model consists of six convolutional layers, followed by two fully connected (FC) layers. Convolutional layers consist of 2D convolutions, batch normalization, and nonlinear activation with a rectified linear unit (ReLU). First and last convolutional layers also employ max pooling with 2×2 kernel to extract more robust features and reduce intermediate output data dimensions. First five convolutional layers operate with 64 5×5 kernels and the last convolutional layer with 12 85×5 kernels. The outputs of the last convolutional layer are then flattened

Table 3 Compositions of training, validation and test sets used to train and evaluate the models

Data set	Weeks	Frames	Proportion
Training	21, 23, 25	7561	46.2%
Validation	24	2938	17.9%
Test	22, 26	5876	35.9%

to a single vector, which is then fed to two 1024 neuron FC layers, both having ReLU activation. Last FC layer outputs the final prediction result. The model was implemented with PyTorch [11] and trained with Skorch [15].

2.4 Training

To gauge the effects of multisource data on the crop yield prediction task with spatial inputs, we performed trainings with four different input data configurations. The data configurations and the input data sources included in them are further given in Table 4. To elaborate, the derived data configurations were as follows:

- *RGB only*. As [9] was conducted with RGB data from UAVs only, we wanted to make baseline performance evaluation with UAV RGB data only. No other sources were included in this setting.
- *No S2*. Next, we wanted to see the effects of soil and weather data on the predictive performance. We thus included all other sources of data (UAV, soil, Veris MSP3, topography, and weather) but excluded the satellite data.
- *S2 Raw*. As Sentinel-S2 Level-2A products contain additional algorithmically generated layers, we wanted to see the effect of including just the raw wavelength bands with other input data sources.
- *S2 Full*. The last setting was to use all data acquired for this study.

Because data were distinct from data used in [9], we initialized and trained all models anew for each data configuration. To account for the effects of randomized network parameter initialization, we trained 10 models per data configuration, 40 trainings in total. We used Adadelta [19] as the optimizer, 0.58 for the learning rate, 0.001 for the weight decay, and 0.9 for Adadelta's ρ coefficient as those were the best performing hyperparameters in [9]. Similarly, we used early stopping with a

Table 4 The different data configurations used for training distinct models. *RGB Only* uses UAV RGB data only. *No S2* uses UAV, soil, Veris MSP3, topography, and weather data. *S2 Raw* adds Sentinel-S2 raw wavelength band data to *No S2*. *S2 Full* adds calculated Sentinel-S2 Level-2A product layers to *S2 Raw*. An X indicates the inclusion of an input data source to a data configuration, while a dash indicates the exclusion

Source	Channels	RGB Only	No S2	S2 Raw	S2 Full
UAV	1–3	X	X	X	X
Soil	23–30	–	X	X	X
Veris	31–36	–	X	X	X
Topo	37	–	X	X	X
Weather	38–39	–	X	X	X
S2 bands	4–15	–	–	X	X
S2 other	16–22	–	–	–	X
Band count		3	20	32	39

patience of 50 stagnant epochs and continued the training once. The models were trained with Nvidia Tesla V100 Volta and Pascal architecture server GPUs in a distributed computation environment.

3 Results

The CNN models with distinct input data configurations were trained with data of four unique fields. The model architectures, hyperparameters, and the training procedures were identical to [9]. As the aim of our study was to evaluate the effects of introducing multisource inputs to crop yield prediction, we trained spatial yield prediction models with four distinct data configurations. The data configurations are discussed in Sect. 2.4. As the training time loss function we used the mean squared error (MSE). Other loss metrics were also calculated, including the square root of the MSE (RMSE), mean absolute error (MAE), mean absolute percentage error (MAPE), and the coefficient of determination (R^2). These metrics (RMSE, MAE, MAPE) were not monitored during training and, thus, did not influence model selection.

The baseline model using UAV RGB data only attained 1055.7 kg/ha test RMSE, 18.2% test MAPE, and 0.343 test R^2. Out of all data configurations, the best performance of 364.1 kg/ha test RMSE, 5.18% test MAPE, and 0.922 test R^2 was achieved using all input data presented in our study (*S2 Full*). The performance results for all data configurations with the held-out test data set are given in Table 5.

To gain a better view into how the models train with distinct data predicted, we also examined the unseen test sample distributions of predicted values against ground truth values, the true crop yields. With the data, the baseline *RGB Only* model's predictions resemble a Gaussian distribution centered around the mean 5140 kg/ha of true yield values. As more inputs are introduced, the predicted distributions' shapes align with the true values more closely, expressing multi-modal peaks where the true values have them. The test set distributions are depicted in Fig. 2.

Table 5 The test set performance of the same CNN architecture and hyperparameter configuration with various data configurations. *RGB Only* is the baseline model. Out of the configurations, the model performed best with all input data layers (*S2 Full*)

Data configuration	Test RMSE (kg/ha)	Test MAE (kg/ha)	Test MAPE (%)	Test R^2
RGB only	1055.7	838.8	18.2	0.343
No S2	892.4	694.9	14.8	0.531
S2 raw	461.0	340.9	6.94	0.875
S2 full	**364.1**	**274.3**	**5.18**	**0.922**

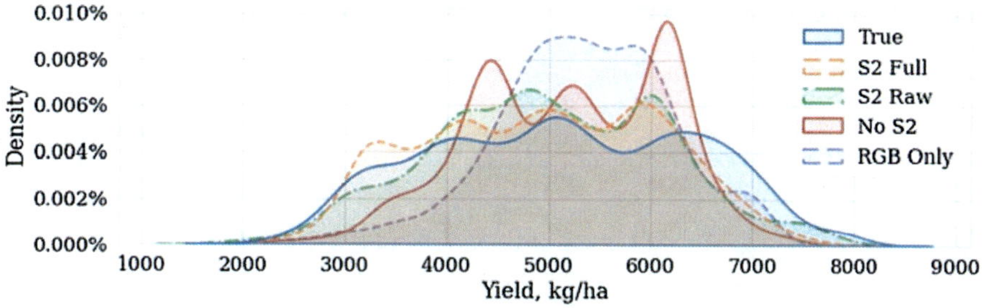

Fig. 2 Distributions of predictions against true yields with the holdout test set

4 Discussion and Conclusions

In this study, we evaluated the effects of using input data from multiple sources on the task of spatial crop yield prediction. Using a CNN model architecture developed for UAV RGB inputs from [9], we introduced additional data from sources like soil samplings, Veris MSP3 soil scanner, topographical maps, weather stations, and Sentinel-S2 satellites to the model. We trained ten models for each distinct input data configuration: (1) a *RGB Only* baseline model, (2) a *No S2* multisource model with satellite data excluded, (3) a *S2 Raw* multisource model with raw satellite band data included, and (4) a *S2 Full* multisource model with all input data. Out of each set of ten trained models, we selected the models performing best. The model architecture and hyperparameters for the training were taken from [9] and left unchanged to constrain the variability in performance to data only. The only thing varying between model trainings, in addition to four distinct input data source configurations, was the initialized model weights.

The performance with a larger number of fields using UAV RGB data has already been extensively studied in our previous studies [9] and [8]. Thus, training a model with only UAV RGB data provides a studied baseline to which models trained with additional data can be compared against. The best performing data configuration was *S2 Full* with 364.1 kg/ha test RMSE, 5.18% test MAPE, and 0.922 test R^2 using all 39 layers of input data for each extracted frame. Compared to the baseline *RGB Only* model, the *S2 Full* attained 65.6% lower RMSE, 67.3% lower MAE, 71.5% better MAPE, and 0.579 higher R^2 with the test set. Generally, every model with multisource inputs performed better than the baseline model. This is shown in Table 6.

Crop yield prediction with spatial data and spatial deep learning models has seen an increase in the past few years [16]. Having been studied with a variety of different architectures, from feed-forward networks to hybrid spatiotemporal models, studies have also been conducted with CNN as the main architecture. In [9], a single CNN model was developed to predict crop yields from fields with varying crop types (wheat, barley, and oat) from UAV images collected from Finnish crop yields during 2017. Using smaller frames extracted from orthoimages, the best performance was

Table 6 The relative performance of the models trained with distinct multisource input data configurations to the baseline *RGB Only* model. Relative improvements in MAE, RMSE and MAPE are given in proportion to the baseline error values. Improvement in R^2 is given in absolute units

	Relative change from RGB only			
Data setting	Test RMSE	Test MAE	Test MAPE	Test R^2
No S2	-15.5%	-17.2%	-18.7%	+0.188
S2 raw	-56.3%	-59.4%	-61.9%	+0.532
S2 full	**-65.6%**	**-67.3%**	**-71.5%**	**+0.579**

484.3 kg/ha MAE and 8.8% MAPE. Using soil nutrient data, seed rate, elevation maps, soil's electroconductivity, and satellite data in USA, the authors of [2] trained a CNN to predict crop yields for nine fields. They report an average scaled MSE of 0.70 which translates to 1145 kg/ha. The authors of [18] utilized RGB and multispectral data acquired with a UAV from rice fields in China to predict rice yields with a composite CNN model on field block scale. Feeding the multisource data to distinct, parallelized CNNs, they report a rice yield prediction performance of 0.50 R^2 and 26.6% MAPE.

As we had sufficient data overlap across multiple input sources and the data were acquired from only four unique fields, objective multisource crop yield prediction performance evaluation requires more care in interpreting the results. Relative increase in performance from best performing UAV data utilizing *RGB only* model to the best *No S2* model with additional soil and weather data was notably small. Largest improvements were gained with the introduction of Sentinel-S2 data. Adding raw Sentinel-S2 bands to the RGB, soil, and weather data increased the performance by 40.8% RMSE, 42.2% MAE, 43.2% MAPE, and 0.344 R^2 from *No S2*. Thus, the increase in performance with Sentinel-S2 is considerably higher than what was achieved with adding soil, topography, and weather data to UAV RGB data.

Data acquisition for remote sensing and multisource input data for smart farming is generally laborious and resource intensive. While satellite data is generated automatically, UAVs require semi-autonomous operation at best and the collection of soil data requires extensive on-site manual labor. With more data from a variety of sources, a more extensive and representative study can be conducted.

Another limitation stems from differences in spatial and temporal dispersion of different input data sources. UAV, Sentinel-S2, and weather data vary temporally in the data we have used, whereas soil samplings, Veris MSP3, and topographical maps do not. As our data was split temporally to training, validation, and test sets, the latter are present in all of these data sets. On the other hand, weather data varies only temporally and constitutes spatial rasters with constant values corresponding to the time of UAV imaging. This means that whether the data is split temporally or spatially, some layer or part of data is always present in training, validation, and test sets. As the authors of [12] point out, deep learning models are able to implicitly learn linear and nonlinear couplings from data with correlations. This means that the deep learning models learn sets of representative features from complex combinations of the inputs and not from single input values on solitude. Furthermore, the performance gains with UAV RGB data combined with temporally invariant soil and ground data is trumped by the performance gains of data configurations using Sentinel-S2 data as additional inputs. This would suggest that the combination of the inputs matters more than presence of distinct, invariant data in training, validation, and test sets. However, the concrete effects of simultaneous layer-level data existence in training, validation, and test data sets are presently unknown to us and, thus, a subject of future research.

Regarding multisource data in the context of smart farming and crop yield estimation, data itself is an evolving research topic. The use of multisource inputs in remote sensing, while focusing on multispectral data acquired from satellite systems orbiting the globe, has been extensively reviewed in [4]. The use of multispectral data from UAVs and the prediction architectures thereof is also a developing topic [6]. Another topic related to spatial data is that of autocorrelation [1]. To address autocorrelation of spatial frames in a future study, the inclusion of pixel-wise location information, as suggested in [1], should be sufficient to inform the deep learning model whether data similarity is due to proximity or some other factors or a combination of them.

In conclusion, our study indicates that increasing the number of input data sources increases the performance of intra-field crop yield prediction. To draw definite conclusions on the most optimal configuration of input data sources, more data is required. With more representative data, generalizable conclusions are more warranted. As the data in this study focuses on a single rowing season, a future plan is to study the generalization of a multisource crop yield prediction model with multiple years of data. Yet in this study, the relative increase from baseline of using UAV RGB only as the input data was notable. Consolidating UAV RGB data with soil and ground topology data already somewhat improves the prediction performance, while largest performance gains were gained from using Sentinel-S2 in addition to UAV RGB, soil sampling, Veris MSP3 soil scanner, weather. and topography data.

Acknowledgments We would like to thank Mtech Digital Solutions Oy for partially funding this research, Tampere University for providing the computational resources, and MIKÄ DATA project for providing us with data.

Conflict of Interest The authors declare that they have no conflict of interest.

References

1. Amgalan, A., Mujica-Parodi, L. R., & Skiena, S. S. (2020). Fast spatial autocorrelation. *Biometrics, 30*(4), 729 (2020). https://doi.org/10.2307/2529248. http://arxiv.org/abs/2010.08676. https://www.jstor.org/stable/2529248?origin=crossref
2. Barbosa, A., Marinho, T., Martin, N., & Hovakimyan, N. (2020). Multi-stream CNN for spatial resource allocation: A crop management application. In *IEEE Computer Society Conference*

on Computer Vision and Pattern Recognition Workshops (Vol. 2020, pp. 258–266). https://doi.org/10.1109/CVPRW50498.2020.00037
3. ESA: Sentinel-2. https://sentinel.esa.int/web/sentinel/missions/sentinel-2
4. Ghamisi, P., Rasti, B., Yokoya, N., Wang, Q., Hofle, B., Bruzzone, L., Bovolo, F., Chi, M., Anders, K., Gloaguen, R., Atkinson, P. M., & Benediktsson, J. A. (2019). Multisource and multitemporal data fusion in remote sensing: A comprehensive review of the state of the art. *IEEE Geoscience and Remote Sensing Magazine, 7*(1), 6–39 (2019). https://doi.org/10.1109/MGRS.2018.2890023
5. Krizhevsky, A., Sutskever, I., Hinton, G.E.: ImageNet classification with deep convolutional neural networks. *Communications of the ACM, 60*(6), 84–90 (2017). https://doi.org/10.1145/3065386. http://dl.acm.org/citation.cfm?doid=3098997.3065386
6. Messina, G., & Modica, G. (2020). Applications of UAV thermal imagery in precision agriculture: State of the art and future research outlook. *Remote Sensing, 12*(9) (2020). https://doi.org/10.3390/RS12091491
7. National Land Survey of Finland: Laser scanning data. http://www.nic.funet.fi/index/geodata/mml/laserkeilaus/mml_laserkeilaus_2016_eng.pdf
8. Nevavuori, P., Narra, N., Linna, P., & Lipping, T. (2020). Crop yield prediction using multitemporal UAV data and spatio-temporal deep learning models. *Remote Sensing, 12*(23), 4000 (2020). https://doi.org/10.3390/rs12234000. https://www.mdpi.com/2072-4292/12/23/4000
9. Nevavuori, P., Narra, N., & Lipping, T. (2019). Crop yield prediction with deep convolutional neural networks. *Computers and Electronics in Agriculture, 163*, 104859. https://doi.org/10.1016/j.compag.2019.104859. https://linkinghub.elsevier.com/retrieve/pii/S0168169919306842
10. PaITuli—Spatial data for research and teaching. https://paituli.csc.fi/download.html
11. Paszke, A., Gross, S., Chintala, S., Chanan, G., Yang, E., DeVito, Z., Lin, Z., Desmaison, A., Antiga, L., & Lerer, A.: Automatic differentiation in PyTorch. In *NIPS-W*.
12. Sun, Y., Guo, G., He, X., & Liu, X. (2019). Multi-level coupling network for Non-IID sequential recommendation. *IEEE Access, 7*(Iid), 186247–186259 (2019). https://doi.org/10.1109/ACCESS.2019.2961182
13. Szegedy, C., Liu, W., Jia, Y., Sermanet, P., Reed, S., Anguelov, D., Erhan, D., Vanhoucke, V., & Rabinovich, A.: Going deeper with convolutions. *Proceedings of the IEEE Computer Society Conference on Computer Vision and Pattern Recognition*. https://doi.org/10.1109/CVPR.2015.7298594
14. Tedesco-Oliveira, D., Pereira da Silva, R., Maldonado, W., & Zerbato, C. (2020). Convolutional neural networks in predicting cotton yield from images of commercial fields. *Computers and Electronics in Agriculture, 171*, 105307. https://doi.org/10.1016/j.compag.2020.105307. https://linkinghub.elsevier.com/retrieve/pii/S0168169919319878
15. Tietz, M., Fan, T. J., Nouri, D., Bossan, B., & Skorch Developers. (2017). skorch: A scikit-learn compatible neural network library that wraps PyTorch. https://skorch.readthedocs.io/en/stable/
16. van Klompenburg, T., Kassahun, A., & Catal, C. (2020). Crop yield prediction using machine learning: A systematic literature review. *Computers and Electronics in Agriculture, 177*, 105709 (2020). https://doi.org/10.1016/j.compag.2020.105709. https://linkinghub.elsevier.com/retrieve/pii/S0168169920302301
17. Warmerdam, F., & Rouault, E. (1998). GDAL—GDAL documentation. https://gdal.org/
18. Yang, Q., Shi, L., Han, J., Zha, Y., Zhu, P.: Deep convolutional neural networks for rice grain yield estimation at the ripening stage using UAV-based remotely sensed images. *Field Crops Research, 235*, 142–153. https://doi.org/10.1016/j.fcr.2019.02.022. https://linkinghub.elsevier.com/retrieve/pii/S037842901831390X
19. Zeiler, M. D. (2012). ADADELTA: An adaptive learning rate method. http://doi.acm.org.ezproxy.lib.ucf.edu/10.1145/1830483.1830503. https://arxiv.org/pdf/1212.5701.pdf. http://arxiv.org/abs/1212.5701

Index

T. Lipping et al. (eds.), *New Developments and Environmental Applications of Drones*,
https://doi.org/10.1007/978-3-030-77860-6

D

E